BROAD AND NARROW INTERPRETATIONS OF PHILOSOPHY OF TECHNOLOGY

PHILOSOPHY AND TECHNOLOGY
VOLUME 7

Series Editor: PAUL T. DURBIN

The titles published in this series are listed at the end of this volume.

OFFICIAL PUBLICATION OF
THE SOCIETY FOR PHILOSOPHY AND TECHNOLOGY

PHILOSOPHY AND TECHNOLOGY
VOLUME 7

BROAD AND NARROW INTERPRETATIONS OF PHILOSOPHY OF TECHNOLOGY

Edited by

PAUL T. DURBIN

University of Delaware

KLUWER ACADEMIC PUBLISHERS
DORDRECHT / BOSTON / LONDON

Library of Congress Cataloging-in-Publication Data

CIP

Philosophy of technology II : broad and narrow interpretations /
edited by Paul T. Durbin.
p. cm. – (Philosophy and technology : v. 7)
Includes bibliographical references.
ISBN-13:978-94-010-6738-6 e-ISBN-13:978-94-009-0557-3
DOI: 10.1007/978-94-009-0557-3
1. Technology–philosophy. I. Series.
T14.P535 1990
601–dc20 90–4068

Published by Kluwer Academic Publishers,
P.O. Box 17, 3300 AA Dordrecht, The Netherlands.

Kluwer Academic Publishers incorporates
the publishing programmes of
D. Reidel, Martinus Nijhoff, Dr W. Junk and MTP Press.

Sold and distributed in the U.S.A. and Canada
by Kluwer Academic Publishers Group,
101 Philip Drive, Norwell, MA 02061, U.S.A.

In all other countries, sold and distributed
by Kluwer Academic Publishers,
P.O. Box 322, 3300 AH Dordrecht, The Netherlands.

printed on acid-free paper

TABLE OF CONTENTS

PART I
THE NATURE OF PHILOSOPHY OF TECHNOLOGY

PART II
DEFICIENCIES IN ENGINEERING ETHICS

PART III
SYSTEMS THEORIES

PART IV
HISTORICAL, CULTURAL, AND POLITICAL CRITIQUES

PREFACE AND ACKNOWLEDGEMENTS

A kind and careful reader who peruses the contents of this volume in the context of its predecessor, Philosophy and Technology, volume 6, *Philosophy of Technology: Practical, Historical and Other Dimensions* (1989), will quickly note the continuity. Both volumes focus on philosophy *of* technology (as contrasted with our customary philosophy *and* technology), and both emphasize diversity and complexity. The introduction to volume 6, by Friedrich Rapp, starts from the premise that the object of academic philosophy of technology is multifaceted – and so deserves multidisciplinary study; this volume begins with the academic beast itself and claims that it should or should not be "specialized" in the customary academic fashion of the day. Two such miscellaneous volumes in a row depart from our plan of alternating conference proceedings with miscellaneous volumes. It seems the departure was worthwhile in this case: the two volumes together (6 and 7) might make an excellent textbook to introduce the student to the marvelous diversity of philosophy of technology today. In any case, the next two volumes in the series (8 and 9) are planned as theme volumes, proceedings based on international conferences; that should correct any imbalance.

Once again, as in volume 6, it seems appropriate to acknowledge with gratitude two sets of people: first, the referees who have evaluated the papers and frequently prodded the authors to make improvements; and second, Mary Imperatore and Dorothy Milsom, who have had to retype so many of the texts.

University of Delaware *Paul T. Durbin*

Paul T. Durbin (ed.), Broad and Narrow Interpretations of Philosophy of Technology, vii.

PAUL T. DURBIN

INTRODUCTION: CONFLICT OVER PHILOSOPHY OF TECHNOLOGY AS AN ACADEMIC FIELD

It is almost two decades since I began editing *Research in Philosophy and Technology*[1] and this Philosophy and Technology series.[2] I have always tried to avoid "philosophy *of* technology" as the name for our fledgling academic subdiscipline. Instead, I have striven to keep the pages of the two series open to all comers, to philosophers of all persuasions (and a handful of non-philosophers) who happen to be united only by a focus on technology or problems of our technological age. But the issue of philosophy of technology as an academic field – however much opposed by practitioners or reviewers of volumes in the two series – will not go away.

For critics, including philosophers who see themselves primarily as sociotechnical critics rather than academics, it remains blessedly true that philosophy of technology is decidedly not yet an academic field in any recognized sense. Many would hope it never becomes such. True, there is our Society for Philosophy and Technology, which runs international meetings of a high caliber every two years. But there is no recognized graduate teaching program in philosophy of technology anywhere in the world; and the number of explicit philosophy of technology courses, undergraduate or graduate, remains quite small. On the other hand, courses in philosophy *and* technology – from engineering ethics to biomedical or computer ethics, from decision theory applied to technological projects to philosophical dimensions of science, technology, and society – are thriving as never before, and often enough in equally thriving programs that include more than philosophers in their makeup.

Under these circumstances, it seems topical once again to raise the question of philosophy of technology as a legitimate or illegitimate claimant to the status of an academic subdiscipline in the cluttered contemporary landscape of fields identifying themselves as such in the modern multiversity. We are fortunate to have here a lead-off symposium that explicitly raises that issue in a sharply-worded confrontation between Joseph Pitt and Frederick Ferré.

Paul T. Durbin (ed.), Broad and Narrow Interpretations of Philosophy of Technology, ix–xvii.

BACKGROUND: DEPARTMENTS, SPECIALIZATION, AND PROFESSIONALIZATION IN AMERICAN HIGHER EDUCATION

For over half of its history, U.S. higher education turned out mostly clergymen and lawyers. Looking back on that period, we might be tempted to think that this meant specialized training for the ministry or the practice of law. That, however, was not the case. What a college education in the U.S. prepared young men (almost exclusively) for, from the founding of Harvard College in 1636 through the founding of hundreds of denominational colleges in the first two-thirds of the nineteenth century, was leadership in the community. Professionalization and specialization only began to take root, and then became the dominant mode in U.S. higher education, in the period roughly from 1860–1920. In subsequent decades, that seemed to many critics to signal the end of what might be called "education in wisdom," the preparation of leaders for a broad range of responsibilities.

Professionalization, specialization, and departmentalization of higher education in the U.S. began in the last quarter of the nineteenth century. According to Alexandra Oleson and John Voss,[3] "The unplanned ... efforts to build a new order of [higher] learning [in the U.S.] were a product of the interaction between European intellectual traditions and the American social and cultural environment."

Though the effort was unplanned, it did have a structure: "The keystone ... was the American university ... [where] without sacrificing the German ideal to advance knowledge for its own sake, American academic reformers ... managed to combine that goal with a broad and direct commitment to what they called 'real life.'"

Between 1870 and 1900, the number of college students in the U.S. increased fourfold from about 50,000 to well over 200,000. (Meanwhile, population in the U.S. doubled from 38 million in 1870 to 76 million in 1900.) Oleson and Voss note that, "In the wave of this expansion came a corresponding demand for teachers trained in the specialized fields of chemistry, physics, modern languages, European history, and even the social sciences, which, under the elective system, had come to form a major part of the new college curriculum." Who was to do this teaching? Where earlier generations of U.S. college teachers had trained mostly in Germany, "By the turn of the century, the number of students enrolled in American graduate programs had risen from fewer than fifty in 1870 to nearly six thousand."

The story of the development of the social sciences – anthropology, economics, political science, psychology, and sociology – is told by Dorothy Ross:[4]

> The tempo of professionalizing activities ... was markedly quickened by the generation of the 1880s and the kind of professionalism they espoused came to dominate their disciplines. There appears to have been at least three professional ideals at work among the early social scientists – that of the practitioner, that of the college based teacher-scholar, and that of the research-oriented scholar-scientist.

Later generations would lament the dominance of research over teaching and service, but in the beginning all three were part of the new professional ideal.

Ross continues: "Professional motivations ... promoted disciplinary unity. Once formed [typically around 1900], professional institutions acted as moderators and neutralizers of ... conflicting views." The point here is that uniformity began to prevail, though everyone recognizes that schools of thought, rather than complete unanimity, dominate the social sciences right down to the present. The organizing key was a devotion to "the advancement of knowledge" in specialist fields.

By the 1920s, this pattern of elective undergraduate education taught by professors trained in graduate programs based in academic departments – with the professors belonging to professional societies dedicated to the advancement of knowledge – prevailed in all the major North American universities. The chief exceptions were religiously-oriented denominational colleges – and even they gave in to the departmental system in short order. Devotees of the generalist approach of the traditional humanities fought against the new compartmentalization of knowledge, but without success.[5]

HISTORY, PHILOSOPHY, AND SOCIOLOGY OF SCIENCE, TECHNOLOGY, AND MEDICINE

The first break in this tradition – with scholars in one discipline studying the work of other disciplines – came with the emergence of specialists in the history of medicine and the history of science.

In the U.S. in the nineteenth century, medical schools were often free-standing, many of them proprietary; by the first decades of the twentieth century, almost all of the leading medical colleges had become units within universities – in some cases, among the most powerful units. Throughout the nineteenth century, similarly, natural philosophy was transformed into science and various science departments found their primary home bases in the universities. The disciplines of history of medicine and history of science followed almost immediately.

From colonial times in the U.S., there had been physicians who wrote on topics in the history of medicine, but the first major history of medicine department was founded at the Johns Hopkins University in Baltimore, Maryland, in 1932. The founder was Henry Sigerist, a Swiss-born graduate of the Institute for the History of Medicine at the University of Leipzig, but it would be decades before the majority of historians of medicine in the U.S. were history-department-trained rather than physician/amateurs.

The story of the professionalization of the history of science followed along somewhat the same lines. Many early historians of science were scientists, science administrators, or science patrons, with professional training in history of science emerging slowly over a period of decades. The leading figure in the early history of the history of science movement in the U.S. was George Sarton, a Belgian-born scientist and gentleman-scholar who founded *Isis* in 1913 and was influential in the founding of the History of Science Society in 1924. Sarton, however, experienced great difficulty in introducing science as a fit object for historians at Harvard University, and the establishment of a history of science department there was long delayed.[6]

The history of philosophy of science in the U.S. exhibits a similar pattern. It was not until emigrants from Austria – most of them members of the Vienna Circle and Jewish refugees fleeing from the Nazis – arrived in the U.S. and established bases in leading universities that the Philosophy of Science Association and its journal, *Philosophy of Science*, came into being. This was in 1934, but the Association did not begin to hold regular professional meetings until 1968. In this case, however, the professional education of philosophers of science began relatively quickly; and they were welcomed without difficulty into already-professionalized philosophy departments in the 1940s and 1950s.

After World War II, other humanistic and social science disciplines studying science, technology, and medicine emerged in rapid succession. *Medical sociology* enjoyed some currency among physicians before, but during the war sociologists were recruited to do work for the military and in the postwar period medical sociology quickly became an accepted specialty in both medical schools and sociology departments. The case was similar with *science policy studies*, based on the experience of the importance of science for military purposes during World War II and the growing importance of science thereafter. The Society for the History of Technology, with its journal, *Technology and Culture*, was founded in 1958. Studies of science by sociologists had appeared in the U.S. in the 1940s and 1950s, but professional training in the *sociology of science* only began in the 1960s and the Society for Social Studies of Science was not founded until 1975.[7]

THE RISE OF INTERDISCIPLINARY PROGRAMS

Representatives of the traditional humanities had objected strenuously to the compartmentalization and professionalization of knowledge in the early decades of the twentieth century. Such protests have continued right down to the present.

After World War II, anti-specialism took on a new and positive coloring. Inter- and multidisciplinary programs came into existence. Good examples are urban studies programs as well as area studies, e.g., American Studies, Latin American Studies, Asian Studies. These were seen by their proponents as forward-looking ventures, as programs needed to better prepare students for a "real world" of problems not arranged along neat disciplinary lines.[8]

Three developments deserve mention in this connection: the rise of programs in bioethics, science, technology, and public policy (STPP), and science, technology, and society (STS).

Bioethics has, within roughly two decades, become an extraordinarily well established field. There is an *Encyclopedia of Bioethics*. There is an impressive, computerized, regularly – updated and federally funded bibliography. There are textbooks and journals and monographs in ever increasing numbers.

What is peculiar about the field of bioethics, aside from the fact that it is multidisciplinary almost by definition, is its anomalous relationship to academia. Of the two leading research centers, the Kennedy Institute and the Hastings Center, one is at a university (Georgetown) and the other is not. Though bioethics courses, and even programs, are emerging with astonishing rapidity, most of these are in medical schools or allied health programs, with almost none in traditional philosophy departments. It is still today a difficult matter to get a doctoral degree in bioethics.

STPP programs, though they are typically multidisciplinary, fall more within the departmental tradition of the early decades of the twentieth century than within the 1970s trend of interdisciplinary programs, including STS. Often championed by scientists or engineers – though typically supported also by discipline-oriented historians, philosophers, and sociologists of science and technology, as well as science policy experts – STPP programs for the most part were designed to train scientific and engineering managers for broader social responsibilities. Up to that time, managers had typically come up through the ranks of researchers or engineering practitioners; now, as the postwar world began to be seen as technically very complex, it seemed that this career ladder was no longer the ideal.

STS programs stand in sharp contrast to STPP programs. In the vast majority of cases, STS professors were trained in the same sorts of graduate programs as STPP professors; that is, they were either scientists and engineers interested in broader issues, or they were trained as historians of science and technology, philosophers of science, etc. The difference is that most STS professors had been strongly influenced by the anti-technology aspects of the student anti-war movement of the late 1960s and early 1970s; and that a great many are also interdisciplinary, even anti-disciplinary, in their orientation. STS professors, much more so than STPP professors, have rebelled against the sort of graduate training they received. Even now, as STS programs take on a less strident anti-establishment coloring, there is still strong resistance to training a cadre of STS professional scholars.

PHILOSOPHY OF TECHNOLOGY IN THIS TRADITION

In almost twenty volumes of *Research in Philosophy & Technology* and *Philosophy and Technology*, there have been a handful of papers expressing extreme anti-academicism. Contributions by and on Jacques Ellul are the most obvious but not the only examples. Joseph Agassi, as one other example, has castigated his academic colleagues and urged them on with strident rhetoric to join in mass movements to save the world from environmental and other catastrophes. But expressions of anti-academicism have definitely been in the minority.

At the opposite extreme, there have been about the same number of authors who represent what Thomas Perry calls "professional philosophy"[9] – philosophy done in a style that consciously apes science and claims to be cumulative in the way science is alleged to be, with neat little articles building on one another in an impressive wall of "advancing knowledge." Joseph Pitt, in his pleas here for a professional philosophy of technology, is not himself a representative of this minority. Most of those who have written in this vein are concerned with such highly technical issues as cost-benefit analysis, and their contributions might just as well have been published in technical journals – even when the authors were critical of the technical fields considered.

Far and away the largest category of contributions to *RPT* and *P and T* have come from academics with much less rigid definitions of what counts as "good philosophy." Indeed, most of them would fall on the "pluralist" side, opposed to "narrow analytical philosophy," in the Great Debate in the

American Philosophical Association of the last ten years.[10] Nonetheless, almost all of our authors are academics in good standing in their departments, and presumably their refereed articles in *RPT* and *P and T* have counted toward their promotions.

A special category here is one that concerns Pitt a great deal – religious commentators on technology and technological society. There have, however, been exceedingly few papers that make explicit appeal to religion in either *RPT* or *P and T* – perhaps five or six. This has not been by design; if presented "respectably," religious contributions are generally recognized as legitimate in academia, and there seems no reason why religion-based contributions ought to be excluded from philosophy of technology.

SOME EDITORIAL CONCLUSIONS

The typical unit of discourse of "professional philosophy" of Perry's extreme sort is the relatively short, journal-length paper that responds to and builds on earlier work in a particular tradition. By contrast, the typical unit of discourse of the extreme anti-academic is a book – anything from a pamphlet-size manifesto to a multi-volume survey of major historical trends.

One might expect, because *RPT* and *P and T* have had something of the character of professional journals, that the overwhelming majority of contributions to the two series would tend toward the professional end of the spectrum. But this has not been so. While today most philosophers not only in the U.S. but in most other countries as well, consider themselves to be professionals in some sense, and while they do tend to write articles more often than books, the modes of their discourse are remarkably diverse even within particular national traditions. It is from this broad mainstream of *loosely* professional philosophers – predominantly from the U.S. but increasingly from a broad range of other countries – that most of our contributions have come. The articles are almost always carefully reasoned, but it would clearly be an exaggeration to say they build on one another in any cumulative way.

To sum up, could philosophy of technology become an academic discipline over the protests both of some leading practitioners and of critics? Possibly. But if the past is prologue, there are likely to be no narrowly "professional" standards. It is more likely to be a loose-knit field, and in the future as in the past the standards employed to decide if an article is publishable are likely to remain extremely diverse.

The present volume is no exception to this rule – as a glance at the table of contents will reveal.

University of Delaware

NOTES

1 The eight volumes I edited have no separate titles, and only volume 7 had a theme, technology and society. Because it will be pertinent at the end of this introduction, I want to comment here on the authors of the essays included in the series. There were just under ninety authors, with approximately a third authoring more than one article. (The most frequent contributor, Carl Mitcham, authored or compiled six essays or bibliographies.) Authors range from the extremely-well-known (in philosophy circles) – such as Joseph Agassi, Mario Bunge, Joseph Margolis, Kai Nielsen, and Marx Wartofsky in the U.S. and Canada, and Jacques Ellul and Friedrich Rapp in Europe – to the very obscure. Many have, during the past fifteen years, become recognized worldwide almost exclusively for their contributions to philosophy of technology: e.g., Albert Borgmann, Don Ihde, Carl Mitcham, Kristin Shrader-Frechette, and Langdon Winner. Others, such as H. Tristram Engelhardt, Jr., have pioneered parallel fields such as philosophy of medicine. What can be said in summary is that they are a remarkably diverse bunch, united only by their conviction that technology is an apt subject for philosophical analysis.

Reference should also be made here to the fact that *Research in Philosophy and Technology* has continued to be published by JAI Press under a new editor, Frederick Ferré. He is editing all theme volumes: Ethics and Technology (guest-edited by Carl Mitcham, 1989), Technology and Religion (forthcoming in 1990), Technology and Politics, and Technology and the Environment.

2 The previous six volumes in the current series are: Paul T. Durbin and Friedrich Rapp, eds., *Philosophy and Technology* 1983 (originally published as volume 80 in Boston Studies in Philosophy of Science series); Carl Mitcham and Alois Huning, eds., *Philosophy and Technology, II: Information Technology and Computers in Theory and Practice* (1986; originally BSPS vol. 90); Paul T. Durbin, ed., *Technology and Responsibility* (1987); Paul T. Durbin, ed., *Technology and Contemporary Life* (1988); Edmund F. Byrne and Joseph C. Pitt, eds., *Technological Transformation: Contextual and Conceptual Implications* (1989); and Paul T. Durbin, ed., *Philosophy of Technology: Practical, Historical and Other Dimensions* (1989). All, of course, were published by Kluwer Academic Publishers.

3 See Alexandra Oleson and John Voss, eds., *The Organization of Knowledge in Modern America, 1860–1920* (Baltimore, Md.: Johns Hopkins University Press, 1979). References in the text are to pp. x–xiii. See also Burton J. Bledstein, *The Culture of Professionalism: The Middle Class and the Development of Higher Education* (New York: Norton, 1976). The next several pages here repeat, in large part, what I have said in "Technology Studies against the Background of Professionalization in American Higher Education," *Technology in Society* 11 (1989): 439–445.

4 Dorothy Ross, "The Development of the Social Sciences," in Oleson and Voss (note

1, above), pp. 107–138. References in the text are to pp. 118 and 122.

[5] See Laurence Veysey, "The Plural Organized Worlds of the Humanities," in Oleson and Voss (note 1, above), pp. 57–106.

[6] See essays by Gert Brieger and Arnold Thackray in Paul T. Durbin, ed., *A Guide to the Culture of Science, Technology, and Medicine* (New York: Free Press, 1980, 1984).

[7] See essays by Linda Aiken and Howard Freeman, Diana Crane, Carroll Pursell, and Jerry Gaston in Durbin, *Guide* (note 4, above).

[8] See especially the articles by Julie Thompson Klein, "The Dialectic and Rhetoric of Disciplinarity and Interdisciplinarity," pp. 85–100, and "The Broad Scope of Interdisciplinarity," pp. 409–424, in Daryl E. Chubin *et al.*, eds., *Interdisciplinary Analysis and Research: Theory and Practice of Problem-Focused Research and Development* (Mt. Airy, Md.: Lomond, 1986).

[9] Thomas D. Perry, *Professional Philosophy: What It Is and Why It Matters* (Dordrecht: Reidel, 1986).

[10] See A.J. Mandt, "The Triumph of Philosophical Pluralism? Notes on the Transformation of Academic Philosophy," *Proceedings and Addresses of the American Philosophical Association* 60 (November 1986): 265–277.

PART I

THE NATURE OF PHILOSOPHY OF TECHNOLOGY

JOSEPH C. PITT

IN SEARCH OF A NEW PROMETHEUS

Toward the end of his book, *Philosophy of Technology,* Frederick Ferré reminds us of that wonderful ancient myth in which Prometheus steals the secret of fire from the gods and gives it to humanity, thereby setting us on the road to civilization. For this Prometheus was punished. He was chained to a rock and tortured daily by a bird who ate his liver, which then regenerated overnight so he could be tortured again by having it re-eaten the next day by the same bird. Not that I would want to deliberately set events in motion which would lead to such a horrible fate for anyone, but I sorely feel the need for a new Prometheus who would set us free from current philosophers of technology whose major objective seems to be to keep informed philosophical discussion of technology out of reach. Such a savior would not have to give us fire; intellectual enlightenment would suffice. As the old Prometheus freed humanity from the torment of the cold and dark, the new Prometheus through his or her gift would free us from the current state of darkness which characterizes the philosophy of technology. Ferré aspires to be that Prometheus, but I find that he is not.

Now these are harsh words, I recognize that. But I must admit to deep disappointment with Ferré's book, and let us be clear from the beginning that my quarrels are over what a book about the philosophy of technology should address. Instead of using the opportunity offered by the Prentice Hall Foundations of Philosophy series to survey the field anew and consider the really exciting developments that have taken place in the last ten to fifteen years, Ferré repeats a tired old song, popular in the earlier part of this century when thinking about technology amounted to little more than indulging in an obscurantist form of metaphysics. We are introduced, or should I say reintroduced, to two alleged "forms of thought" in human affairs, practical versus theoretical, which are then reified so as to make it possible to generalize about the effects of these kinds of thinking. Philosophical questions about technology are then characterized by way of presenting bright or somber visions. Ferré's models for these visions are such minor philosophical giants as Karl Marx, Buckminster Fuller, Herbert Marcuse, and the ever present

Paul T. Durbin (ed.), Broad and Narrow Interpretations of Philosophy of Technology, 3–15.

Martin Heidegger. I say "ever present," meaning by that "present in a specific form of philosophical thinking about technology," one obsessed by high flying questions about the significance of modern life, man's relation with god, etc.

Now I admit to a certain prejudice here. Or perhaps it is a matter of philosophical style, training, or intuition. Whichever, and I am sure it is a combination of all of these, the question I am concerned to address is this: to what extent does Ferré's book give us an honest feel for the nature of philosophical problems raised by and peculiar to technology as currently constituting the foundations of philosophical thinking about technology? Since I do not think it does much in that direction, it is incumbent on me to explain why.

To begin with, philosophical questions of interest concerning technology center around the decisions we make regarding technological enterprises. By "technological enterprises" I mean such diverse projects as the building of a nuclear plant, the development of a new weapons system, the creation of new consumer goods and their marketing, the creation of social institutions and the development of knowledge. The main problems concern the nature of the knowledge and the values we use to make decisions about the development and implementation of projects in these areas. In these contexts major questions arise concerning the kinds of risks which might be incurred and the reliability of techniques for correctly estimating those risks as well as other consequences, intended and unintended. While he sometimes nods in some of these directions, it is clear that the kinds of concerns just listed are not those of Ferré. To see the distance between us let us take a quick review of the contents of his book by chapter.

According to Ferré, "This book aims to clarify the foundations of the philosophy of technology" (p. 7). Toward determining whether or not Ferré achieves that end, consider first the chapter titles: I. What is the Philosophy of Technology?; II. Defining Technology; III. Technology and Practical Intelligence; IV. Technology and Theoretical Intelligence; V. Technology and Modern Existence; VI. Ethics, Assessment and Technology; VII. Technology and Religion; VIII. Technology and Metaphysics. Not only do the chapter titles fail to give one a sense of organic unity in the text, but some, particularly V, VII and VIII, actually jar the senses. Let me take my worries one at a time.

First, with regard to the worry about lack of unity in the book. This charge is both fair and not. It is fair in the sense that from the titles alone one has no sense of where the author is going. Nor is this feeling of confusion alleviated when you actually read what Ferré has to say. For example, given the structure of chapter I, in which we are essentially given a brief taxonomy of the classical areas of philosophy and an attempt to fit technology into each of these, one is left with the expectation that the rest of the book will elaborate these relations. Not so, or at least not obviously so. For example, no further effort is made to talk about technology and epistemology, at least not in any clear sense of epistemology. More importantly there is actually no mention of the history of philosophy as one of the classic areas. This may be because Ferré does not see how it bears on technology. But since he virtually ignores the question of the relationships between science and technology and the growth of science and its influence on the development of epistemology, it should come as no surprise that the history of philosophy/natural philosophy would be ignored.

Now Ferré might object that I am not being fair. After all, there are two chapters which are at least implicitly epistemological in character, namely three and four, the two dealing with theoretical and practical intelligence. These chapters arise out of Ferré's definition of technology in chapter 2, i.e., the practical implementation of intelligence. I have few difficulties with the definition. It resembles in many ways Emmanuel Mesthene's characterization of technology as "the organization of knowledge for the achieving of practical purposes" (Mesthene, 1970, p. 25) or my own view that technology is humanity at work (Pitt, 1983, 1988b). But then to distinguish between theoretical and practical intelligence using Alfred North Whitehead as the primary philosophical source is not very helpful. In the first place, the material of Whitehead's that Ferré uses is from the 1920s. The world of the 1920s is not the world of today. Thus, there is an important sense in which Whitehead's naive and simpleminded view of both science and technology fails to have any relevance for the kinds of concerns Ferré raises later when questions about assessment and man's existence in the *modern* world are considered. There is also a second reason why Whitehead's distinction is not very helpful. The Whiteheadian distinction is itself, even for its own time, worthless. All one has to do is study the history of science and technology taken together to consider the important interrelations among these areas, and one finds that there is no theoretical intelligence *simpliciter,* just as there is no practical intelligence pure and simple.

Now for some it may be out of order to suggest that philosophical theses can be shown to be false by appeal to facts. But philosophy is not an apriori activity. If the results of philosophical work are to be taken seriously, then they must reflect the ways things are. To assume that there exists a kind of thinking basic to technology which is different from some other form of thinking, without justifying that assumption by appeal to what actually happens in the world of science and technology, seems to me to weaken the philosophical enterprise as well as to produce a framework which is misleading.

Let me be even more aggressive here. I am dismayed by Ferré's sense of what an appropriate historical source is and how such sources are to be used. Whitehead, whatever insights he may have had as a logician and metaphysician, is certainly not an historian. Even if he were, surely it is incumbent upon Ferré to use the best and most up-to-date historical work to ground his philosophical theories. But more to the point, speaking as a Galileo scholar myself, Whitehead is no Galileo scholar. And yet, using Whitehead's romantic, rationalistic, and poetic version of history, Ferré proceeds to draw some rather strong conclusions. Let us look at this in some detail.

After talking about theoretical reasoning as having been made possible by the invention of logic, Ferré adds critical evaluation to its powers. He then looks at some of the strengths and weaknesses of theoretical reason. The first point concerns the intrinsic value of theoretical reason. After a quick look at Socrates he concludes with the following:

> The passion of reason for its own dispassionate pursuit has erupted many times in history since Socrates. An especially poignant element in the struggle between Galileo and the authorities of his day was the sense of the intrinsic importance of the unfettered thought. Despite threats of torture, public humiliations, and house arrest, Galileo continued to think his theories, and the world of thought was changed (pp. 42–43).

What does it mean to say that despite adversity Galileo continued to think and the world was changed? Let us look at how Galileo affected the world. René Descartes, upon hearing of Galileo's trial, wrote Pierre Mersenne that he would not publish his *Le Monde*. Yes that changed the world, but not in the way Ferré seems to suggest. Second, Galileo is well known as a physicist. What is not well known is that Galileo's physics died when he did. There is a good reason for this. Galileo was not a particularly good mathematician. His physics is couched in the language of geometry – a language and a method that, even at the time he was working, was being superseded by dramatic new developments in mathematics due to Johannes Kepler, Bonaventura Cavalieri, and finally Isaac Newton.

What thoughts did Galileo think and the world changed? Perhaps Ferré is referring to Galileo's alleged defense of Copernicanism in his *Dialogue on the Two Chief World Systems*. Well, the *Dialogue* certainly brought Galileo notoriety. But, as I have argued elsewhere, Galileo's "defense" of Copernicanism is far less than a defense and much more than it has been made out to be (Pitt, 1988a). When Galileo wrote the *Dialogue* he essentially seized an opportunity offered by the Church to do one thing and used it for his own ends. The Church was concerned to show the world that its ban on efforts to reconcile Nicholas Copernicus's views and the Bible was not based on ignorance of either the available facts or the theories in question. As Italy's foremost scientist and a loyal Catholic, Galileo was asked to take on that job. But the Church was unaware of Galileo's own agenda. He had been working on a treatise on the tides since 1615, based on a theory he most probably stole from a friend as far back as 1595. What Galileo tried to do in the *Dialogue* was to propose his theory of the tides as evidence which could decide between the two competing theories of Copernicus and Ptolemy/Aristotle. In fact, the *original* title of the so-called *Dialogue on the Two Chief Systems of the World* was *Dialogue on the Tides*. Galileo was forced to change it simply to *Dialogue* by the censors and it was not until the eighteenth century that it came to be called by the name we now associate with it. Yes, by all means let us praise theoretical reason, if there is such a thing. But let us also remember that the history of the creation, development, and defense of ideas is not the history of a metaphysical fiction called theoretical reason doggedly pursuing its own ideals for its own sake. It is a complicated story tied up with practical matters, historical contingencies, serendipity, and luck. But to remember this requires that we have a healthy appreciation for history in the first place and, in the second place, that we not allow ourselves to be lulled by the poetry of metaphysics into believing in "disembodied intelligences" (p. 40).

Let us turn to Ferré's second case of the strengths and weaknesses of theoretical reason. This involves reason's attention to detail, which for Ferré can be both a good thing and a bad thing. It is bad when it results in something like scholasticism. Quoting again from Ferré:

> Socrates inspired Plato, who taught Aristotle the importance of careful distinctions. Centuries later the Aristotelian tradition of discriminating discriminable ideas was in full flower among the scholastics of the Middle Ages. The opponents of Galileo were not irrationalists: they were solidly grounded in the vast achievements of theoretical intelligence gained by their predecessors. The revolt of Galileo was not against unreason but against what Whitehead calls the "inflexible rationality of Medieval thought" (p. 43).

This passage especially points up the problem of relying on Whitehead as an historical authority. Whitehead and Ferré make it sound like there was one thing called medieval thought. There was not. Consider some well known facts about the period from 1250 to 1550. In the thirteenth century, the European intellectual world, which was dominated by fragments of the work of Plato, was introduced to the work of Aristotle. It caused such an uproar in Paris, where the material arrived *via* Spain and the Arabic tradition, that it caused fights in the streets and the eventual closing of the University of Paris for a period of ten years. That was when Aristotelian thinking entered the West and Western thought. Prior to that time there was no Aristotelian tradition in the West. Furthermore, what became available to Paris and the young Thomas Aquinas was Aristotle in Latin translation from the Arabic. It was not until a hundred years later that Agostino Nifo finally translated Aristotle into Latin from the Greek. In the sixteenth century a new translation of Archimedes was produced in Spain; although it is in conflict with some of the works of Aristotle, it developed a following of its own. In the fifteenth century we witness the rediscovery of the classical Roman authors who are responsible for the development of humanism, yet another tradition in conflict with some of the followers of Aristotle. Where is the inflexible rationality of medieval thought?

I am pushing this line of attack because I am worried about the way in which Ferré develops the views he is advocating. If his sources and his historical claims are not reliable, what should we conclude about the strength of his arguments?

I began this particular line of inquiry in the course of looking at the structure of Ferré's book. I suggested that if one looked for coherence in the book by looking at the chapter titles one could not find it. I then began to look at the content of the chapters. So far we have noted Ferré's initial attempt to link technology to the general areas of philosophy, having left out history of philosophy as a general area; and we have seen the consequences of ignoring history for Ferré's claims about reason. We have also looked briefly at his distinction between practical and theoretical reason, a distinction derived from metaphysical speculation and devoid of content. But these worries pale before my total dismay at his selection of Marx, Marcuse, Buckminster Fuller, and Heidegger as representatives of various perspectives within the area Ferré marks out as technology and modern existence. As I suggested earlier, these individuals do not impress me as philosophers of technology. Marx *is* an important figure, primarily in terms of his historical importance in world political events. To the extent he has views on technology, those views are actually part of his account of political economy. Ferré's failure to place Marx's views properly makes it difficult to use them as Ferré wishes, namely as an indicator of the "breadth of possibilities" (p. 74) available for critique. Marx's concern with technology as a factor in a treatise on political economy reduces our understanding of the breadth of Marx's vision and is, therefore, at odds with Ferré's announced intent.

Buckminster Fuller is Ferré's second figure to be considered for the breadth of possibilities. I find Fuller without merit. One might look to him as a visionary or a poet or for metaphysical inspiration, but the muddled metaphysics he brings to his efforts to be reflective about technology hardly assist Ferré's efforts to clarify the foundations of the philosophy of technology.

Marcuse is of marginal interest, and only to those of us within a certain generational profile and most probably because of social and political events in the sixties; his philosophical contribution will not survive the decade, much less the century.

As for Heidegger, the fourth and final thinker Ferré appeals to to help us broaden our possibilities, well there is a Heidegger industry. It is extremely strong in Europe and within the phenomenological school of American thought. But of what value is it when we come to questions of modern existence? Heidegger is not of our age or our culture. He belongs to a time gone by and was confused by the transformations it was experiencing – more confused even than we are today. In an attempt to overcome that confusion, Heidegger invented a language and a philosophical methodology which is obscure and idiosyncratic, making his work impenetrable, which is the opposite of mind expanding.

These comments are concerned to worry a single point. By relying on dubious sources and by appealing to obscure and – from the viewpoint of the *philosophy* of technology – irrelevant views, how does Ferré accomplish his announced task of clarifying the foundations of the philosophy of technology? The aim of clarification is not to be achieved by laying out a range of confused and incomprehensible views in the name of mental enlargement. Clarification requires careful explication and examination by example. And even if this were done, how such an examination of a truncated Marx, a befuddled Fuller, an outdated Marcuse, and an impenetrable Heidegger serves to clarify the foundations of the philosophy of technology is not in the least bit obvious.

The entire question of which philosophers to use as exemplars in a discussion of technology and modern existence might be answered by simply dropping the topic. What kind of a philosophical question is being asked here? Of what value is a bright or a somber view of technology without in-depth case studies of particular problems associated with discrete technologies and technological developments? Finally, is it really fair to those also interested in the overall impact of technology on the development of modern society to use a set of psychological categories like bright and somber to clarify a philosophical problem, assuming there is one?

But more important, what does it mean to address a topic such as "technology and modern society"? The way Ferré discusses this issue leads us to assume there is some one thing out there called Technology. The big "it." But what is the justification for this assumption? When discussing Marx, Ferré refers to the "technological phenomenon" – note the singular. Fuller I will ignore despite the fact that Ferré continues to talk about technology as if it were an it in that context. Heidegger rumbles on about the essence of technology – and if there is an essence to it there must be the thing which has the essence. When discussing Marcuse, Ferré uses such terms as "the new technological repression" – without details and particulars. What is this "it" that represses? What do we gain by reifying technology this way? The only gain I can see is the ability to render grand, sweeping claims about grand, sweeping issues, none of which serve to clarify that which Ferré wishes to clarify. You do not clarify the foundations of the philosophy of technology by assuming the very thing under examination, the nature of technology. In short the reification of technology is misleading and fails to clarify anything.

Earlier I claimed that I see the philosophically interesting questions about technology to concern the decisions we make about technological enterprises. The main problems are the nature of the knowledge and the values we use to make decisions about the development and implementation of technological projects. Chapter 6 of Ferré's *Philosophy of Technology* is titled "Ethics, Assessment and Technology," and it might be thought that it would be concerned with material appropriate to the project of clarifying the foundations of the philosophy of technology. Unfortunately it is not. The chapter begins with the following statement, "Before proposing any ethical critique of modern technologies or raising specific questions for ethical discussion, we need to review our available resources and pause to remind ourselves of what such a critique can and cannot hope to accomplish" (p. 75). My question is, why is Ferré setting the discussion in this ethical framework? If the objective is to clarify the foundations of the philosophy of technology, why are we now embarking on an ethical critique of technology? Ostensibly the reason is to contrast ethical critiques with other sorts of assessment, such as cost-benefit analysis. But how does this sort of contrast of types of assessment clarify the philosophical foundations of technology?

What Ferré actually gives us is a truncated and rather simple-minded theory of value (resulting in such profound principles of evaluation as "Do not destroy good," "Try to create good," and "Be fair") followed by an attack on cost-benefit analysis, followed by a list of ethical problems for technologies. There is nothing wrong with examining ethical issues raised by technological advances and innovations. It is however deeply misleading and far from clarifying to *assume* that ethical critiques of technology are superior to cost-benefit analyses, especially when such critiques are based on such a meager theory of value. The situation here is further aggravated by the series of non-sequiturs Ferré uses in place of full scale arguments. Before taking a look at some of Ferré's claims about cost-benefit analysis, it should be noted that cost-benefit analysis has been criticized by others (see Shrader-Frechette, 1985). It is not a perfect method and has a number of problems, some of which Ferré correctly identifies, even when his arguments are less than convincing.

Ferré's first objection to cost-benefit analysis concerns the difficulty of placing numerical values on ethical considerations. This is certainly a problem, but no more so than trying to decide between any two goods when only one can be achieved and each is of sufficient importance not to obviously outrank the other. In short, it is not a problem peculiar to cost-benefit analysis, although Ferré would have us think that to be the case. But, what is worse is that from his worries here he derives a totally unwarranted conclusion. "Quantification and monetization of goods and harms in cost- and risk-benefit analysis tend, accordingly, to *bias decisions toward present exploitation of opportunities and against the values of conservation for the future*" (p. 82). Aside from the fact that no argument is given for this claim, it is false. Sliding over the complexity of the issues involved here, Ferré has conveniently ignored the difference between short term and long term considerations and he has not allowed that decision-makers can discuss such differences reasonably, which they can and do. Likewise with his claims about the value of human life. Why are decision-makers who employ some quantitative methods less able to make value judgments about human life than anyone else? The answer is not given, but the claim is nonetheless made.

The difficulty with Ferré's account of cost-benefit analysis, and his later discussion of the inability of such analyses to apply something called the Principle of Justice, is that he attempts to leave the reader with the conclusion that only formal quantitative approaches to assessment are fundamentally flawed. Honesty, if nothing else, requires that we note that all the ethical systems in the world taken together have not corrected the same problems

Ferré faults cost-benefit analysis for. This is not to defend cost-benefit analysis. As noted earlier, some of the problems Ferré points to are legitimate ones. What these objections are designed to point out is that Ferré has not presented the case for these various types of assessment in a manner which clarifies their roles. Properly restricted, cost-benefit analysis is concerned with predictions of outcomes given a certain knowledge base. It can even be used to rank order outcomes in terms of preference *if* the values employed in the analysis can have their rank order and relative importance agreed upon, which is not to say that it is possible. Ethical assessments of technological projects are not concerned with the knowledge base for the projections, except to point out that it is always insufficient for any completely accurate judgment. The role of ethical assessment is to raise questions about the quality of life and the overall character of the change such projects can introduce into social systems. There is a proper role for each. But they do not operate at the same level (see Pitt, forthcoming).

Now I would like to turn to the final two chapters of Ferré's *Philosophy of Technology* and ask why *they* are there. In the chapter on Technology and Religion we find Ferré addressing such earthshaking topics as Technology and Original Sin and "the tool and the tao." Frankly I am bewildered. If his objective here is merely to rehearse a variety of religious attitudes toward technology, that is all well and good but does not belong in a book that aims to clarify the foundations of the philosophy of technology. That is, unless we have a profound confusion here between philosophy and theology. This I suspect may in fact be part of the problem. For it is clear that religious concerns overwhelm Ferré's account throughout the book. We note this from the beginning when the story of Abraham and Isaac is used to exemplify issues in axiology. Shortly thereafter we find Ferré talking about how axiological questions in metaphysics would raise questions about god. The chapter, Ethics, Assessment and Technology, is replete with biblical examples, something I found particularly offensive. We have spent several hundred years working to disentangle legitimate questions of moral concern from the perspective of religious dogmatism. I see no reason to confuse the two now.

Finally we turn to Ferré's last chapter for which the topic is technology and metaphysics. The topic seems to follow from Ferré's use of such expressions as "the reality of human freedom in the face of apparent technological determinism." See what happens when you reify technology – now we have technological determinism! So, clearly, when faced with determinism we must talk about human freedom, a classical topic in metaphysics.

In short, the topic and the discussion is consistent with the position that accepts the reification of technology and the grand view of history. But where is the evidence for technological determinism? How does the introduction of such assumptions clarify the foundations of our thought on these matters? In point of fact, the indiscriminate introduction of such unchallenged assumptions has just the opposite effect. It not only fails to clarify, but locks the character of our thinking about technology into a rigid and unproductive framework. Unfortunately, Ferré once again returns to Galileo – and since we have already spent some time on Ferré's problems with characterizing Galileo clearly, let us forget about metaphysics, freedom, and mechanistic models of man and see what we can do by way of summary.

How can you write a book about the philosophy of technology and not discuss in detail either engineering or science? How can you write a book about the philosophy of technology and worry primarily about religion and metaphysics? These questions have troubled me deeply after having studied Ferré's *Philosophy of Technology* with some care. The answer to these questions requires that we step back from the specifics of my worries and try to consider the problem in a more general mode. The issue between philosophers who fall into Ferré's camp and those who might agree with me resembles the kind of issue that has for some years divided historians of science. For some time now there has been some question as to whether one should concentrate on the specific activities of a single scientist, reading his working notes and his correspondence and then trying to reconstruct his scientific biography, or whether one should broaden one's perspective and consider the social climate and the forces acting within it on the individual in question to explain why he or she did what they did. That is, do we try to explain the scientist's behavior by examining only what he said and did or do we look to social factors for that explanation? The battle here is between internalists and externalists, with the externalists being the social factor types.

In an earlier version of these comments given at the symposium on Ferré's book sponsored jointly by the Society for Philosophy and Technology and the Society for Process Philosophy in Washington, D.C., in December, 1988, I proposed that we see the two major divisions in the discussion about how to do the philosophy of technology as reflecting a similar difference regarding scope and method. I argued as follows:

> What Ferré wants to do is look at questions about technology from an intellectually global perspective. He sees questions about technology as having importance beyond the immediate problem of a polluted river. For him, and for the externalists in general, we cannot isolate technology from the overwhelming fact of the human condition.

Thus, we need to address the problems of technology from that perspective. To do this means we push the horizon back from the petty problems of daily existence to questions about the overall fate of humanity. It is in this context that we can at least see how or why one might want to address the problem of the global relation between technology as the product of man's activity and religion as the product of man's fear of the unknown.

But to see technology from these Olympian heights does not mean that we have exhausted the philosophical problems it presents. There now, you have me doing it – talking about technology as an it! There are also philosophical questions that require an internalist's perspective. By this I mean, there are questions that legitimately arise when we consider specific technological activities. For example, when attempting to calculate the probability that a given means of disposing of nuclear waste will protect the environment, one has to ask questions about the relation of evidence to hypothesis, about the character of the knowledge claim that is being made, as well as about the morality or value of disposing of waste in that way. In fact, before one can ask the moral question, one has to ask questions about the concept of confirmation being used and about the concept of knowledge being employed. Is knowledge about technological phenomena different from knowledge about other sorts of things? And the answer to this question from an internalist's point of view cannot proceed from some apriori conception of technology written large because that would beg the question.

Now just as it ought to be clear that one cannot write adequate history of science by ignoring either the scientists or the circumstances in which he or she worked, it seems clear that an adequate philosophy of technology has to take account of both externalist and internalist concerns. In conclusion, then, my major quarrel with Ferré comes from his failure to acknowledge the internalist program as a philosophically legitimate component of the philosophy of technology. Perhaps now we can work toward a means of achieving such a goal.

Ferré rejected this proposal, asserting the necessity for metaphysics but not explaining how that clarifies the foundations of the philosophy of technology. In place of my previous proposal, which I now too reject – because through Ferré's response it was obvious that in matters of intellectual honesty one can give no quarter when faced with dogma – I now suggest the following. If philosophy and philosophers are to be of value we must talk to the world in which we live. That means our account of technology must be sensitive to the very field Ferré left out of his taxonomy of philosophical fields, history; otherwise how can we claim to know what it is or how it came to be that way? Our critical views on technology must be informed by as complete a knowledge of its history as possible including its impact on the various aspects of human society. We must come to technology through history and only then can we ask the important philosophical questions concerning the nature of technological knowledge and worry about the proper means of assessing technological projects. An historical approach to understanding the roles technology has played in human evolution will force us to abandon talk about Technology as a single thing and will expose the bankruptcy of such notions as technological determinism. Only then will we see that just as the criteria for scientific knowledge have changed over time, so too has our understanding of what constitutes technological knowledge. We will see that attempts to box technology into a single kind of thing fail as new social technologies emerge with their own set of problems and concerns. To be of philosophical value we must abandon apriori metaphysical characterizations of technology and anything else, stop relying on superstitious religious values, and get to work learning about the thing we wish to understand. This is not to deny that we come to the study of anything with certain assumptions already in place. But the key to good scholarship and learning is that we allow those assumptions to be changed in the face of new knowledge. In short, it is time we undertook the clarification of the foundations of the philosophy of technology *in context, where it means something.*

Virginia Polytechnic Institute and State University

REFERENCES

Ferré, Frederick. *Philosophy of Technology* (Englewood Cliffs, N.J.: Prentice-Hall, 1988).

Mesthene, Emmanuel G. *Technological Change* (New York: Mentor, 1970).

Pitt, Joseph C. "The Epistemological Engine," *Philosophica* **31** (1983): 77–97.

Pitt, Joseph C. "Friends in Holy Places," *The Sciences,* January/February 1988[a], pp. 50–56.

Pitt, Joseph C. "'Style' and Technology," *Technology in Society,* **10** (1988[b]):447–456.

Pitt, Joseph C. "Technology and the Objectivity of Values," in L. Embrey, ed., *Technology and the Life World* (Lanham, Md.: University Press of America, forthcoming).

Shrader-Frechette, Kristin S. *Risk Analysis and Scientific Method* (Dordrecht: Reidel, 1985).

FREDERICK FERRÉ

DEFINING HORIZONS: A REPLY TO JOSEPH C. PITT

Joseph C. Pitt has written "harsh words," as he recognizes, against my book and against the entire philosophic enterprise it represents. Since fire must sometimes be fought with fire, I suppose my response will also sound harsh. This is not my accustomed tone, but for the issues between us to stand out as sharply as their importance requires, such straight talk from me is unavoidable.

I

First I must reject Pitt's bizarre characterization of my purposes in writing this book. He portrays me as seeking the role of a new Prometheus and then, having chained me to that rock, he sets to ripping at my entrails. Thanks for the honor, but no thanks. My purposes were much less titanic.

What I was invited to do by Prentice-Hall was to write a concise textbook in the Foundations of Philosophy series, designed to be useful in many classroom contexts. Like all the other titles in the Foundations of Philosophy series, *Philosophy of Technology* is meant to be used in connection with other books. Sometimes it will be used in advanced philosophy courses to tie issues together that will then be developed in greater depth; sometimes it will be used in introductory philosophy courses, perhaps with other Foundations of Philosophy texts, to lay out one topic among a series of topics; sometimes it will be used in general interdisciplinary courses, like "Western Civilization" or "Issues of the Modern World"; and sometimes it will be used outside the liberal arts altogether, as in alert schools of engineering (like the two where I have myself taught), where faculty and students try to get a handle for reflection on what they are doing.

Preparing a book for such varied uses is a difficult assignment to write to, but worthwhile. It requires the author to aim at the high points, remaining clear and accessible but always being much more succinct than anyone might wish. This motivates an author to reflect on and display the essentials. That is just what I tried to do, and it turns out that my choice of essentials is exactly

Paul T. Durbin (ed.), Broad and Narrow Interpretations of Philosophy of Technology, 17–23.

what Joe Pitt so strongly dislikes. What he opposes is my decision to try to show that philosophy of technology has matured beyond a collection of specialized sub-disciplines and now can take its place within the mainstream of philosophic interests.

This latter was the other primary purpose of the book. It is not only a textbook offered for use by other educators, it is also an extended argument for the integrated reconsideration of philosophy of technology, to be treated like any other "philosophy of..." topic, by other philosophers. Perhaps in attempting this I was, after all, taking up an immodest, if not Promethean, task. But it seems that the time is right for insisting that the urgent questions raised by technology can be illuminatingly set into the horizon-defining philosophical topics of methodology, epistemology, axiology, and metaphysics. In aiming at this I never considered that I was stealing anything from anyone but, rather, adding something useful: not fire from heaven, perhaps, but a little light from the philosophical tradition.

In this context it strikes me as strange that Pitt complains that I "neglect" the history of philosophy. On the contrary, my aim was to bring the full history of philosophy into living contact with philosophy of technology. History of philosophy is of course not parallel, logically, with the four great philosophical domains of methodology, epistemology, axiology, and metaphysics. All four have their own histories of development. Inquiries in all four domains can be pursued systematically, not only historically, which is why there can be *philosophy of history* as well as *history of philosophy*. But to this topic I shall return at the end.

II

Before then I need to expose and rebut a few of the worst examples of Pitt's inaccurate allegations about my text. He has removed from his present printed version the specific, glaring errors I pointed out at our December encounter, but unfortunately there was no practical way then (nor is there even now) for me to be comprehensive. Therefore his next layer of outrageous statements remains to be challenged. For the reader's convenience, I shall take up his assertions not in any logical order but in the order of his somewhat jumbled presentation.

(1) Pitt suggests that my own way of characterizing philosophical questions about technology is mainly in terms of "bright or somber visions" (this volume, p. 10) of technological influences on society. This is not the case. The chapter that offends Pitt, presenting four dialectically interrelated

positions on the overall social impact of technology, is not even introduced until the half-way point of the book. By then the methodological and epistemological chapters have shown how I prefer to characterize philosophical questions of those sorts; and by then I am about to turn to axiology. Chapter 5 – which begins with Karl Marx, summarizes the views of Buckminster Fuller and Martin Heidegger, and ends with Herbert Marcuse (who, as a Marxist Heideggerian with Fuller-like confidence in the powers of technology, illustrates the possibilities for tightly integrating apparently conflicting extremes) – is presented without any critique, positive or negative, from the author. These ways are not my ways of doing philosophy of technology, but they have been important ways for many people in recent decades; a book purporting to provide the highlights of our field for multiple purposes would be deficient if this approach were not even presented. Indeed, in my own classroom work in this subject I have found that students are roused to vigorous discussion by considering these alternatives, and also that they discover after a while that these "visions" are not enough to rest on. This, as well as accurate reportage, was the point of providing the chapter. Considering such large, conflicting views not only adds breadth of possible viewpoints but also helps convince students of the need for more precise methods of assessment and critique. This is what I tried to say in the last paragraph of that chapter:

> Without alternative visions, our minds tend to be locked, without our knowing it, within narrower outlooks based on more limited experience. The assessment of a great issue needs to begin with a recognition of the spread of choices that can and must be made. Then, when our minds have been opened to the breadth of possibilities, we are better prepared for the careful process of critique (*Philosophy of Technology*, p. 74).

Unfortunately, Pitt does not seem to value this sort of opening for his students' minds, or – by the look of it – for anyone else's. I hope other readers will have a more favorable response to what I was up to in that chapter.

(2) Pitt claims that I merely "nod" in the direction of the decisions we make about the development and implementation of technological projects. He says that "it is clear that the kinds of concerns just listed are not those of Ferré" (p. 4). But how can he justly say this when I devote the longest chapter in the book exclusively to such concerns? True, in a book of this kind I had to settle everywhere for a sharply curtailed discussion. But that curtailment is required of all the Foundations of Philosophy books. In the context of that constraint, I offered a short but essential outline of an ethical theory, a

succinct critique of traditional cost-benefit thinking, and an application of ethical assessment in exemplary, discussion-starting ways to specific technological issues like automation, computers, nuclear energy, technology transfer to developing countries, and genetic engineering. That hardly strikes me as a "nod." Was it Pitt's attention that nodded at this point?

(3) Pitt attempts to show "lack of unity" (p. 4) in the book by listing the chapter titles and professing bewilderment. If he had read the first chapter more closely, he would have realized that the plan of the book had been explicitly stated:

The plan of this book can now be summarized in light of all that has been said so far. Chapter 2 will continue the present focus on *methodology* and critical clarification, both on the definition of technology itself and on the logical character of our job in doing that defining. Chapters 3 and 4 explore aspects of *epistemology* raised by technology insofar as modern technology rests on types of knowledge, both practical and theoretical. Chapter 5 presents four general visions of distinctively modern technology, as we exist within it. Chapters 6 and 7 then pursue principally issues in *axiology*, discussing the implications of technology for ethics and religion. Finally, Chapter 8 carries certain of these issues to the level of *metaphysics*, on which, with a speculative look to the future, the book concludes (*Philosophy of Technology*, pp. 12–13).

There are, I submit, few books that declare their plan and structure so plainly and so early. How could Pitt claim bafflement over the book's organization after that?

(4) Even more astonishing is Pitt's related claim that after the first taxonomical chapter "no further effort is made to talk about technology and epistemology, at least in any clear sense of epistemology" (p. 5). How can anyone who has read with any attention the two chapters (3 and 4) dealing with practical and theoretical knowledge, their contrasting natural domains, their competing standards of success, their intimate interconnections and mutual influences, and their final unity, make such a statement? Perhaps Pitt would appeal to his "at least in any clear sense" escape clause. But since he has not himself given any alternative "clear sense" of epistemology, we are unable to understand his complaint.

(5) In the same astonishing paragraph, Pitt claims further that the book "virtually ignores the question of the relationships between science and technology..." (p. 5). This frankly brings me from astonishment to stupefaction, since the main point of chapters 3 and 4 was, at bottom, to illuminate the mutual repulsions and attractions between science and technology. I admit that I tried to go about this while avoiding the hackneyed clichés that Pitt may have been expecting. But I explain this tactic clearly:

Some readers may have wondered why there have to this point been so few uses of the word "science." This has been no accident. It is too easy to say that "science" is the secret of the vast development of technology in the modern world. It is also too simple to say that technology is the key to modern science. Both statements are partially true and partially false. Modern science, as it has developed since the seventeenth century, is the joint product of theoretical and practical intelligence, and so is modern technology. Neither gave birth to the other. They are non-identical twins of the same parents (*Philosophy of Technology*, p. 44).

The position announced in that paragraph is repeatedly defended. I find it extremely difficult to understand how Pitt could have missed it.

(6) Pitt's next claim is that I hold a position exactly the opposite of the one I explicitly argue for. He seems to think he is saying something contrary to my (and Whitehead's) position when he announces that students of history will find "that there is no theoretical intelligence *simpliciter*, just as there is not practical intelligence pure and simple" (p. 5). Exactly! That is one of Whitehead's theses in *The Function of Reason* and it is mine also in this book. For purposes of preliminary analysis it may be illuminating to note the internal oppositions between intelligence used for practical purposes and intelligence used for theoretical goals; but at last one must acknowledge that intelligence is a unified function. That was indeed the concluding theme of the two epistemological chapters, which I stated bluntly in the opening paragraph of the section entitled (in bold face) "Intelligence United":

Despite the usefulness of our key distinctions between practical and theoretical intelligence and between technology and science, we must not forget that in the real world these distinguished aspects of reality are not finally separated. They are interactive in various ways, and in the end their interactions are the most important thing about them (*Philosophy of Technology*, p. 48).

(7) Two more brief rebuttals should be enough. Pitt says (p. 11) that I "assume" that ethical critiques of technology are superior to cost- and risk-benefit analysis. No. I *argue* for this superiority. I patiently show how quantification, monetization, "shadow marketing," etc., leave vital matters out of account or lead into conceptual problems. Similarly, he says that "no argument" is given that the practice of quantification and monetization of goods and harms tends to bias decisions in favor of near-term exploitation and away from conservation. But in the very paragraph from which Pitt quotes my conclusion, an argument was clearly given, showing how in a monetized context interest rates, linked to discount rates, lead to the discounting of future values in direct proportion to the interest rate. Given *any* interest rate, the bias is to realize potential returns before future dollars drop to

negligible values. Pitt says that my claim is false. I doubt that; but, be that as it may, what is demonstrably false is his statement that I offer no argument.

III

Let me now move from particular defenses and counterattacks to a short consideration of the real battle between Pitt and me. This has less to do with his sloppy reading of my book than with, as he says, "what a book about the philosophy of technology should address" (p. 3). It has to do with our approaches to philosophy itself.

My approach to philosophy is open to a variety of efforts and methods. My own favorite philosophers are Socrates, Wittgenstein, and Whitehead. The one thing I insist on is that philosophical thinking must be *both* critical (intelligence self-controlled by method which we call logic) *and* situated within the comprehensive horizons of questions that break through every effort to make them narrow. I have given my arguments for this in the first chapter of my book and will not repeat them here. Not every philosophical issue needs to be itself comprehensive in this sense, but every issue, if it is raised philosophically, needs to be able to be related to such comprehensive horizons as: "What is it to be known at all?" "What is it to be valuable at all?" "What is it to be real at all?"

Pitt, at the December meeting, suggested a "compromise" in which, as I interpreted it, the critical side of philosophy of technology could go off to do its thing unencumbered by the comprehensive side. I rejected that as a "separate but equal" proposal that would, if accepted, mean the death of real philosophy of technology. Thinking does not qualify as philosophy just by being comprehensive; it has to be critical in the highest degree. But thinking does not become philosophy just by being critical; it is distinguished as philosophy from other critical fields by the inescapably comprehensive contexts of its questions. To isolate the critical from the comprehensive as a temporary matter, perhaps while at work on a particular problem, is perfectly legitimate as long as it is recognized for the expedient it is – an acknowledgment of human cognitive finitude – but a permanent Jim Crow society of philosophers of technology would be intolerable.

Now Pitt has thought better of his proposal. Instead, he is offering another way of defining the horizon for philosophy of technology, not through the traditional issues of philosophy but through history. As I said earlier, I have no quarrel with history. Its study is also comprehensive in its own way. Everything in time can be treated historically. And the historical way of defining the horizon for a subject can be wonderfully illuminating. History may not be entirely free of *a priori* elements, as Pitt seems to think, but it can be used in ways, as we have seen in recent revolutions in philosophy of science, that can challenge old dogmas in salubrious ways.

Still, philosophy of technology must not be satisfied to let its final horizons be defined by history. The essential questions of philosophy will not be side-stepped. They have their own history and they transcend that history. "What is historical knowledge?" "How does that kind of knowledge relate to other sorts of knowledge?" "Are values simply historical?" "What are the values we seek in studying history?" "Are any values permanent?" "Is there anything real outside of time?" "Is the past in any sense real?" "Does the future somehow 'tenselessly' exist or is it being constantly created by present events?"

Pitt has come a constructive distance from his first proposal. He no longer recommends that philosophers of technology settle for split-brain vision. He now urges us to search for historical coherences. If he keeps moving toward these wider horizons he may yet discover philosophy.

University of Georgia

PETER LIMPER

PROCESS THEMES IN FREDERICK FERRÉ'S *PHILOSOPHY OF TECHNOLOGY*

One of the distinctive features of Frederick Ferré's recent book, *Philosophy of Technology (PT)*,[1] is its affinity with process thought and its use of some specific concepts from Whitehead's philosophy. In the following discussion I shall focus on an explication and critical examination of this aspect of Ferré's book.

I should note at the outset that this book does not present a systematic and detailed exposition of Ferré's own views. In a rather brief compass he attempts to introduce the reader both to philosophy and the philosophy of technology, and to summarize the ideas of a number of representative thinkers concerning technology. My account of what I take to be Ferré's position on the topics I discuss will necessarily involve a certain amount of interpretation and "reading between the lines."

ULYSSES AND PLATO

The influence of process thought is most obvious in Ferré's third and fourth chapters, entitled "Technology and Practical Intelligence" and "Technology and Theoretical Intelligence." Here he draws heavily upon Whitehead's discussion, in *The Function of Reason (FR)*,[2] of the distinction between "practical" and "speculative" reason, the "Reason of Ulysses" and the "Reason of Plato." Not only the distinction itself but much of the characterization of the two types of reason is directly based on Whitehead's account. Practical reason or intelligence is said to be interested rather than impartial, concerned only with what works and with a sufficiently detailed understanding to "get the job done." It "knows no other standard than success," and is "often 'tacit' about its premises and assumptions" (*PT*, pp. 39–40; cf. pp. 34–35). Its value is instrumental, rather than intrinsic (*PT*, p. 35). Theoretical intelligence, on the other hand, is said to be "(a) disinterested, (b) curious about what is so and why methods work, (c) concerned for detail, (d) highly self-critical, and (e) explicit" (*PT*, pp. 38–39). It has intrinsic value, and "its own aesthetic as well as ethical imperatives" (*PT*, p. 35).

Paul T. Durbin (ed.), Broad and Narrow Interpretations of Philosophy of Technology, 25–35.

Ferré has already defined technology as "practical implementations of intelligence" (*PT*, p. 26). Thus the reader may be tempted at first to assume that practical intelligence alone is to be associated with technology, and perhaps that speculative or theoretical intelligence is to be identified at least in part with science. Ferré's account is more subtle than this, however. He recognizes (as does Whitehead) the complex interplay between theory and practice which is characteristic of *both* modern science and modern technology. Modern technology has "outfitted Plato with the tools of Ulysses," the technological "instruments and apparatus" that have characterized scientific research from the time of Galileo (*PT*, pp. 44–46). At the same time, science has "expanded Ulysses' vision with the perspective of Plato" in modern science-based technology (*PT*, pp. 46–48). The result has been the union of theory and practice in what Ferré calls "techno-scientific thinking," which dominates contemporary life (*PT*, pp. 49–53).

A number of questions can be raised concerning the above account. I will focus on three. (1) Is the role of theoretical intelligence in technology *overstated* in Ferré's discussion? (2) Is theoretical intelligence always "disinterested" and "explicit," or should it be? (3) Can human intelligence always be characterized as *either* practical or theoretical in Ferré's (and Whitehead's) sense, or is there some *other* type of "intelligence" which is particularly important in thinking about the social implications of technology?

(1) Philosophers dealing with technology from several different perspectives might question the real importance of theoretical intelligence in technological society. As Ferré himself points out:

> A strong theme in Marx's thought ... is the primacy in human life and history of the Reason of Ulysses. ... For the real driving force in history, look to the material challenges and conditions of life, particularly to the universal needs of living things to provide for their own subsistence. These needs and our methods of meeting them are in human terms the economic realities. It is these realities which stimulate the responses of Ulysses. They are what actually rule the world, however much the fact may be disguised by the glib-tongued Plato (*PT*, p. 55).

From the standpoint of pragmatism, John Dewey has also argued that in the modern period "the method of changing the world through action" has taken primacy over the older "quest for certainty" which characterizes the tradition of purely theoretical reflection.[3] Finally, we may note the phenomenological analysis of Don Ihde in "The Historical-Ontological Priority of Technology over Science." Ihde argues for a "materialist" interpretation of science and technology which holds that modern science depends not only on technological instruments but on a "technologically mediated basic experience" and a "basic praxis within a life world, a praxis which inclines or predisposes us toward what becomes a scientific world view."[4] Here again, the practical perspective is seen as not only prior to but in at least one sense more fundamental than the theoretical.

All of the above accounts of the relative roles of the practical and theoretical are themselves open to debate. Whitehead, for example, would respond to Dewey with the assertion that pragmatism takes insufficient account of the fact that "there is a strong moral intuition that speculative understanding for its own sake is one of the ultimate elements in the good life" (*FR,* p. 30; see *PT,* p. 35). The point is, however, that when Ferré follows Whitehead in making a fairly sharp distinction between two "functions of reason," and in characterizing modern technological thinking as "theory based practical intelligence" (*PT,* p. 38), he may be begging an important question concerning the role of purely theoretical or speculative reasoning in the development of modern technology.

(2) From the three philosophical standpoints just mentioned, questions might also be raised concerning the "disinterested" nature of theoretical reasoning. A Marxist critique would claim to show that such reasoning always serves a hidden class interest. Pragmatists and phenomenologists would argue that all reasoning is "interested" in another sense, since it is grounded in some concrete human concern.

The claim that theoretical intelligence should aim for total explicitness of understanding might also be seen as problematic. Ferré acknowledges that "highly intelligent activities, like flying an airplane, may be better done without excessive explicitness of the principles embodied in the activity," but asserts that "for theoretical intelligence ... there seems to be no limit to the benefits of explicitness" (*PT,* p. 34). An important recent challenge to this view comes from some critics of artificial intelligence (AI) research, with its assumption that human reason can be completely formalized on the model of a computer program. Joseph Weizenbaum, for example, bases his critique of AI in part on the assertion that the most fundamental human wisdom can never be fully articulated; he quotes with approval Ionesco's statement that "not everything is unsayable in words, only the living truth."[5] Hubert Dreyfus (along with his brother, Stuart Dreyfus) has also argued that, unlike computers, human "experts" do not follow rules that either are or could be fully explicit. He sees this specifically as a challenge to the "reason of Plato," that is, to the philosophical tradition begun by Plato, who "followed Socrates in thinking of wisdom as whatever could be spelled out in explicit principles or definitions."[6]

The important issue here, I think, is not whether *some* sorts of thinking exemplify "the reason of Plato" as articulated by Ferré and Whitehead; clearly they do. As Dreyfus notes, mathematical thinking is paradigmatic of this sort of reason. The more fundamental question is whether Plato and Ulysses between them exhaust the alternatives. This brings me to my third point.

(3) Both Ferré and Whitehead are rather hard on Ulysses. As Ferré remarks, "He does not ask what is fair to the Cyclops or what precedents for the long run he may be setting with Circe" (*PT*, p. 39). And Whitehead states that,

> Some of the major disasters of mankind have been produced by the narrowness of men with a good methodology. Ulysses has no use for Plato, and the bones of his companions are strewn on many a reef and many an isle (*FR*, p. 8).

Practical reason is "narrow," interested in success at any price, and often guilty of "obscurantism" in resisting novelty and the questioning of established methods (*PT*, pp. 37–38; *FR*, p. 34). Yet does this mean only that it needs correcting by what Ferré refers to as the "gadfly sting" of theoretical or speculative reason? A critic might respond to Whitehead by noting that some of the major disasters of mankind have also been produced by the blindness or fanaticism of those who are totally committed to a particular theory. Is there some other option?

In the discussion previously cited, Dreyfus suggests that we attend to another tradition in philosophy exemplified by Aristotle, who

> pointed out that underlying all explicit knowledge of the sort Plato championed there had to be a kind of judgment that enabled experienced practitioners in any domain to apply their principles to particular cases. So Aristotle said, if you have an ethical problem, do not ask a philosopher what to do; ask a wise old man.[7]

The question I wish to raise, then, is whether we need to consider something like Aristotle's idea of political and ethical judgment as an alternative both to the narrow practicality of Ulysses and the speculative flights of Plato. In Aristotle's own terms, must we not acknowledge *praxis* as well as *technē* and *theoria*?

Much recent philosophical discussion of technology has dealt with the role of ethical and social *praxis* in a technological society. A number of positions can be distinguished. At one extreme, what might be called the "technocratic" view asserts that *praxis* should be subsumed under or replaced by appropriate technique; the problems of modern society should be dealt with by experts using a "technology of human behavior."[8] Others argue that technique itself is or ought to be informed by social and ethical thinking; in the words of Paul

Goodman, "technology is a branch of moral philosophy."[9] A Marxist version of this position would be that with the overcoming of the exploitation and alienation of the present economic system it will be possible for technology to take its rightful place in a truly meaningful human *praxis*.[10] Others have held that what Ferré calls "techno-scientific thinking" differs in kind from true ethical/political judgment, and see the problem of our age as one of recovering a genuine social *praxis* in the face of the increasing dominance of narrowly instrumental or technological thinking.[11] An even stronger view holds that social *praxis* as well as the practice of technology is fatally infected by the desire for power and domination; only by turning away from action to a purely non-manipulative aesthetic contemplation can we escape the increasing problems of modernity.[12]

Once again, my point here is not to try to decide between these various perspectives but to suggest that they call attention to an important issue, one which Ferré seems to pass over by his use of Whitehead's distinction between theoretical and practical. He might have found elsewhere in Whitehead's writings (e.g. in *Adventures of Ideas*)[13] a richer account of the range of human thought and its relation to action. However, in neglecting the question of whether there is a distinctive form of social/ethical "intelligence" and suggesting that it is to *theoretical* reason that we should turn to correct the excesses of technology, Ferré may be reflecting a certain intellectualist bias which is characteristic of Whitehead and much of process thought. I shall return to this poinow.

METAPHYSICS: MODELS, VALUES, AND SOCIAL IMPLICATIONS

At the end of his chapter on technology and theoretical intelligence, Ferré asserts that contemporary "techno-scientific thinking" is "open to debate" in two important respects: "the quest of theoretical intelligence for *coherence*" and "the demand that theory be *adequate* to all the data" (*PT*, p. 52). Although these issues are raised in the context of a discussion of the "epistemological authority of techno-scientific thinking," Ferré clearly sees them as having metaphysical implications, and he returns to them in his concluding chapter on technology and metaphysics. Here he expands on his critique as follows.

> First, by the incoherence of modern techno-scientific thought is meant its drive to specialization and analysis, splitting up problems for piece-meal attack rather than weaving solutions together into differentiated wholes. ... Second, by the inadequacy of techno-scientific thinking is meant its tendency to ignore or even deny non-

quantifiable aspects of experience and to force the reduction of subtle intuitions, like personal freedom..., into the inhospitable framework of metaphysical models that allow them no room. From such methods of thought come modern technologies insensitive to the vital non-quantitative requirements of human life, like beauty and justice (*PT*, pp. 131–132).

Having thus diagnosed what he sees as the fundamental flaws of techno-scientific thinking, Ferré makes tentative proposals for an alternative "post-modern" framework.

He states that, "One suggestion well worth exploring is that we try to draw our metaphysical models from ecology,... rather than from the physics-based technological models of the modern world" (*PT*, p. 132). He argues that ecology necessarily "recognizes everything as connected to everything else," thus countering the techno-scientific tendency to incoherence, and that it "cannot do its scientific work without acknowledging the rich spontaneity, the purposes and values, of the living world which is its subject matter," and therefore is "free to be *adequate* in framing its concepts and its models" (*PT*, p. 132). In the concluding paragraph of his discussion of "alternative metaphysics," Ferré explicitly associates such a metaphysics with process thought.

In such ways, it may be speculated, a significantly different mode of thinking, a full post-modern metaphysical scheme allied to post-modern forms of science, may be a candidate to replace modern techno-scientific thinking with a new form of intelligence. The form would be more coherent, more adequate, and more oriented to life. This is high-flying speculation. Whitehead and his followers have done much to spell out such a "philosophy of organism," as Whitehead called it; but... little embodiment of the scheme has made its way into contemporary institutions (*PT*, p. 133).

Although Ferré's discussion here is brief and tentative, I believe that it does reflect his own preference for a "post-modern metaphysical scheme" allied in some way with process thought.[14] In discussing this view, I will not directly question his assumption (which in many ways I share) that such a scheme would be superior in important ways to the implicit metaphysics underlying the contemporary techno-scientific worldview. Rather I will raise certain questions and problems concerning the viability of such a scheme in the current context.

(1) Ferré's suggestion that we "draw our metaphysical models from ecology" is certainly consonant with his advocacy of a process perspective. Although Whitehead did not (to my knowledge) make use of the term "ecology," he could obviously be called an ecological thinker in his stress on the mutual relations at all levels of organisms and environment. However,

Ferré himself provides a cautionary note when he points out that "even ecology itself may eventually become a domain for reducers and analyzers" (*PT*, p. 134). Certainly the tendency in much of biological science today can be seen as one of reduction and analysis; one thinks, for example, of the project to map the human genome and to associate particular physical and even behavioral characteristics with specific aspects of human DNA. Furthermore, as Ferré notes in an earlier chapter, developments in biotechnology and genetic engineering raise questions about the extension of the reductive and manipulative approach of modern technology to life itself (*PT*, p. 95). All in all, it seems to be an open question whether a non-reductive ecological perspective will retain sufficient significance to serve as a model for an organic metaphysics.

More broadly, Ferré's own very interesting discussion of the important role of models in metaphysical thinking (*PT*, pp. 121–123) brings to mind the question of whether the modern "techno-scientific" landscape provides *any* readily accessible models for a process metaphysics. Whitehead himself drew on concepts from quantum mechanics and relativity in support of his metaphysical vision (see, e.g., *Science and the Modern World,*[15] chapters 6–8). It is my impression that developments in physics since his time support or at least do not contradict that vision. However, neither quantum physics nor the devices based on it (such as the transistor) are well understood even by most educated individuals today; the more "visible" and comprehensible technologies of the modern world (such as the airplane) still exemplify principles of classical mechanism. Again, as Ferré points out, "the dominant technology for current imagination is ... the electronic computer" (*PT*, pp. 127–128), particularly as it suggests a model for understanding the human mind. Certainly the concept of mind as a rule-based information processing system does not seem to cohere well with the Whiteheadian notion of it as a sequence of living occasions of experience. Furthermore, as Ferré notes, the computer model of mind seems to have no place for genuine freedom or creativity, which in Whitehead's view are fundamental aspects not only of human mind but of the "actual occasions" which are constitutive of all of reality.[16]

It seems that contemporary science and technology offer few models that unambiguously support a Whiteheadian "philosophy of organism." Perhaps this is one reason for the fact, noted by Ferré, that such a philosophy has found "little embodiment in contemporary institutions," if indeed the availability in ordinary thought and experience of appropriate models for a metaphysical system has something to do with its wide acceptance. On the other hand, it is not obvious that this must be the case, or that, even in a technological age, such models must be drawn from science and technology. Also, both the future course of many scientific and technical developments and their possible impact on philosophical thinking are uncertain. I would suggest that there is a challenge and an opportunity for philosophers sympathetic to process thought to seek out and articulate contemporary models which support their metaphysical perspective.

(2) Ferré states that one measure of the "inadequacy" of "techno-scientific thinking" is its "insensitivity" to such value concepts as "beauty and justice." With this Whitehead would certainly agree. In a discussion of the beginnings of the industrial era in *Science and the Modern World,* he commented on the bad effects of what he would later call the metaphysical fallacy of "vacuous actuality."

> The assumption of the bare valuelessness of mere matter led to a lack of reverence in the treatment of natural or artistic beauty. Just when the urbanization of the western world was entering upon its state of rapid development, and when the most delicate, anxious consideration of the aesthetic qualities of the new material environment was requisite, the doctrine of the irrelevance of such ideas was at its height (*SMW,* p. 281).

For Whitehead, of course, beauty or aesthetic quality is the fundamental value concept. Certainly any process metaphysics would be sensitive to the importance of beauty and critical of the continuing disregard of aesthetic quality as a significant value in technological society. Rather surprisingly, Ferré's own discussion of "intrinsic" or "nonmoral" goods does not draw on process concepts of value but is based on Abraham Maslow's account of the hierarchy of human needs (*PT,* pp. 76–77).

In the passage just quoted, Whitehead also speaks of the "limited moral outlook" of the nineteenth century and suggests that it too is rooted in an inadequate metaphysics.

> The doctrine of minds, as independent substances, leads directly not merely to private worlds of experience, but to private worlds of morals. ... Accordingly, self-respect, and the making the most of your own individual opportunities, together constituted the efficient morality of the leaders among the industrialists of that period (*SMW,* p. 281).

By implication, abandonment of the idea of independent substances (the "fallacy of simple location") in favor of a Whiteheadian organic philosophy would help to overcome the "limited moral outlook" of individual self-interest, and presumably support a more just society.

It is not clear, however, how much help there is to be found in

Whiteheadian philosophy in dealing with *specific* questions of justice, particularly as these arise in relation to modern technology. Ferré notes three issues which are often ignored in conventional technological decision making: the need to be concerned for *distributive justice,* for *intergenerational justice,* and for *ecological justice* (*PT,* pp. 84–85). Let us consider the last point first. Ferré notes that, "Our views on how much within nature deserves to be given ethical respect will depend on our metaphysics" (*PT,* p. 85). Now it is often pointed out that a Whiteheadian metaphysics, with its concept of value as pervading all of nature, does in fact provide a basis for a concern for "ecological justice." The problem I see is that it is very difficult to make the transition from Whitehead's view that *all* nature is composed of occasions of experience which involve some realization of intrinsic (aesthetic) value to any particular claim about the relative worth of such occasions or the fairest way for such values to be apportioned.[17] There has certainly been some significant work done on applying process thought to environmental ethics,[18] but I think that much more needs to be done in this respect. More generally, I would argue that process philosophy needs to develop some better way to connect its broad metaphysical concepts with specific ethical principles if it is to be useful in dealing with any of the concerns for justice which Ferré has outlined.

(3) The passages which I have cited from *Science and the Modern World* suggest Whitehead's view that the misuse of technology has been in large part the result of a misguided metaphysics, or at least that the fundamental philosophical errors which he associates with much of modern science have infected technological society as well. Presumably, then, if we were to get our philosophy right (specifically if we were to recover an understanding of reality which recognized the pervasive presence of value and the interconnectedness of things) we might get our technology right as well. "The moral of the tale," says Whitehead at the end of *Science and the Modern World,* "is the power of reason" (*SMW,* p. 299). As Robert Neville puts it, Whitehead's "overall moral vision was the Whiggish one that civilization progresses as ideas become embodied in institutions."[19] In his call for a "post-modern metaphysical scheme" to aid in rethinking and reshaping technology, Ferré seems to share this vision.

Today, however, there are certainly many to question such Whiggish perspectives. Langdon Winner, for example, sounds a skeptical note concerning the relevance of metaphysics for the transformation of modern technological society when he argues that the expectation that a rapid shift of attitudes could follow a philosophical critique is ... badly misplaced. ...To have completed even the most effective critique of Western metaphysics is no guarantee that one will know what, if anything, to do next.[20]

Winner asks us to seek a middle ground between metaphysical speculation and narrow practical concerns: a ground occupied in part by political theory.[21] While I do not altogether agree with his rather scornful dismissal of the possibility of a "new ethic" based on a new metaphysical vision, I believe that his own work on the political aspects of technology raises very important issues. I think that philosophers of technology (including process philosophers) need to attend not only to the way in which our metaphysical theories may influence our technical and social practice, but also the way in which our practice may affect our attitudes toward reality and value.[22]

Discussion of technology as a *political* phenomenon seems notably lacking in Ferré's book. For example, he attributes the tendency of present technologies to "perpetuate themselves" to the conservatism of practical intelligence (*PT*, pp. 37, 131), and does not deal with the possible role of vested interests or political mystification. Perhaps this again reflects the Whiteheadian influence despite Whitehead's own experiences in political life. Process philosophy as such has offered little in the way of political analysis and critique of modern institutions. Once again, I raise this point not so much as a criticism but as a challenge. Can process philosophers contribute to the development of a coherent "post-modern philosophy" which offers both breadth of speculative vision and tools for concrete ethical and political thinking? Certainly I think that this is needed if those like Ferré (and myself) who are sympathetic to the process perspective are to come to grips with the important issues which technology poses for contemporary philosophers.

Christian Brothers College

NOTES

1 Frederick Ferré, *Philosophy of Technology* (Englewood Cliffs: Prentice Hall, 1988). Later references in the text are cited as *PT*, with page numbers.

2 Alfred North Whitehead, *The Function of Reason* (Princeton: Princeton University Press, 1929). Later references, *FR*, with page numbers.

3 John Dewey, *The Quest for Certainty* (New York: Putnam, 1960), p. 3 and *passim*.

4 Don Ihde, "The Historical-Ontological Priority of Technology over Science," in *Existential Technics* (Albany: State University of New York Press, 1983), pp. 25–46.

[5] Joseph Weizenbaum, *Computer Power and Human Reason* (San Francisco: Freeman, 1976), p. 201.

[6] Hubert Dreyfus and Stuart Dreyfus, *Mind over Machine* (New York: Free Press, 1986), p. 2.

[7] Dreyfus and Dreyfus, *Mind over Machine,* p. 2.

[8] See, e.g., B.F. Skinner, *Beyond Freedom and Dignity* (New York: Knopf, 1971).

[9] Paul Goodman, "Can Technology Be Humane?" reprinted in Albert Teich, ed., *Technology and the Future* (New York: St. Martin's Press, 1986), p. 275.

[10] See, e.g., Bernard Gendron, *Technology and the Human Condition* (New York: St. Martin's Press, 1977), chapters 11–14.

[11] See, e.g., Richard Bernstein's discussion of Hannah Arendt in *Beyond Objectivism and Relativism* (Philadelphia: University of Pennsylvania Press, 1983), pp. 207–223.

[12] See, e.g., David L. Hall, *The Uncertain Phoenix* (New York: Fordham University Press, 1982), chapter 5. I think that this position also has something in common with that of Heidegger's later work.

[13] Alfred North Whitehead, *Adventures of Ideas* (New York: Macmillan, 1933).

[14] See Frederick Ferré, *Shaping the Future: Resources for the Post-Modern World* (New York: Harper and Row, 1976), especially pp. 96–100, for an earlier discussion of the relevance of process thought for a "post-modern" vision.

[15] Alfred North Whitehead, *Science and the Modern World* (New York: Macmillan, 1926). Later references, *SMW,* with page numbers.

[16] Recently the "conventional" computer model of the mind as a deterministic rule-based information processing system has been challenged by the "connectionist" model. See, e.g., William Bechtel, "Connectionism and the Philosophy of Mind: An Overview," *Southern Journal of Philosophy* **26,** Supplement (1988):17–41. It remains to be seen whether connectionism or other new developments in computer science will provide models of mind which make room for freedom and creativity in Whitehead's sense.

[17] This issue is discussed at greater length in my paper, "Technology and Value: A Process Perspective," *Contemporary Philosophy* **12** (January 1989):21–23.

[18] See, e.g., Charles Birch and John Cobb, Jr., *The Liberation of Life* (Cambridge: Cambridge University Press, 1981).

[19] Robert Neville, *The Puritan Smile* (Albany: State University of New York Press, 1987), p. 82.

[20] Langdon Winner, *Autonomous Technology* (Cambridge, Mass.: The MIT Press, 1977), pp. 132–133.

[21] Winner, *Autonomous Technology,* p. 134.

[22] I should stress that in raising this question I am not casting my lot with those who see the "end of metaphysics" or even the "end of philosophy" in the post-modern period.

FREDERICK FERRÉ

CLARIFYING AND APPLYING INTELLIGENCE: A REPLY TO PETER LIMPER

Peter Limper raises three important questions, all centered around the character of intelligence and its roles in theoretical and practical pursuits. He also poses some significant challenges for future philosophical work that needs to be done. I am grateful for his perceptive reading and acknowledge that when he read "between the lines," as he said, he invariably got it right.

That leaves me only the task of responding to his three questions and reinforcing his call for serious new efforts from process philosophers and others interested in nurturing constructive political change.

I

Starting, for convenience of exposition, with Limper's second question, "concerning the 'disinterested' nature of theoretical reasoning" (this volume, p. 27), I hasten to acknowledge my long-standing conviction that total disinterestedness in human affairs, including theoretical affairs, is a fiction. It is a limit concept, an abstraction not met with in life. It is, however, an important *ideal* fiction. That is, as an *aim* for theoretical intelligence, the qualities of disinterestedness, the avoidance of rationalization based on special pleading or hidden agendas, is an essential virtue. Theoretical intelligence is permeated by values, of course, but among them (enhanced and reinforced by training) are the austere values of the conscious subordination of personal interests and outlooks to wider considerations of evidence and more general perspectives.

Explicitness, too, is something that we can never totally achieve in human thinking. Weizenbaum, as quoted by Limper, is doubtless correct. But the aims of theoretical intelligence *qua* theoretical call for as much explicitness as possible. All of Limper's appeals to the essentially inexplicit in "wisdom" and "living truth" and the procedures of "experts" are drawn from domains of application which go beyond merely theoretical motivations and rewards. This gives them a weight and importance that sometimes outstrips even the

Paul T. Durbin (ed.), Broad and Narrow Interpretations of Philosophy of Technology, 37–41.

importance peculiar to theorizing. Theorizing, however, has its own intrinsic importance and has gradually evolved its own exacting standards, of which full explicitness is one, and disinterestedness is another.

II

This brings me to Limper's next question, "whether Plato and Ulysses between them exhaust the alternatives" (this volume, p. 28). My answer, in the book and again here, is emphatically not. In the book I began by discussing the common features of intelligence (*Philosophy of Technology,* pp. 30–33), only later exploring the tensions between the needs of intelligence in pursuit of practical ends and the needs of intelligence aimed at theoretical ends. When Whitehead's figurative term for practical intelligence, the "Reason of Ulysses," was introduced, I was careful to state that I was naming *a capacity* and *an aspect* of intelligence. When I later introduced the "Reason of Plato," I took pains to warn that this may be "a useful figurative way of speaking, but 'it' is not a thing." The English language generously allows us to use the word "it" to refer to any contextually defined subject-matter, whether a "thing" or not, like the weather or a university or scientific method or (once defined) technology or intelligence.

Just as I began my discussion of intelligence as a unified function rooted in biology, so I ended with a section entitled "Intelligence United" (*Philosophy of Technology,* pp. 48–49) summarizing the many intricate ways in which the interplay between the sometimes opposing functions and aims of intelligence – dealt with in the intervening two chapters – create unified patterns of self-stimulation and self-control. I hope that these reminders reassure Limper not only that I agree with his sense that there is indeed "some other option" than choice between practical and theoretical intelligence but also that in fact I am not invidiously "hard on Ulysses" (this volume, p. 28). I mean to be hard on each, if not corrected by the other. "Ulysses" *alone,* taken as the ideal type of practical cleverness without any of the breadth of vision provided by "Plato," has, in my opinion, all those defects Limper quoted me as charging against it; but, likewise, the ideal type of "Plato" *alone,* tending (as Whitehead said) to "charlatanism," is not particularly commendable. The point is that they are useful fictions, employed to clarify the important tensions at work – and in need of counterbalancing – in real, complex human thinking.

III

If my answers to these two questions are clear, Limper's opening question, to which he returns in another form at the end of his paper, can now be more readily addressed. His question is of "the real importance of theoretical intelligence in technological society" (this volume, p. 26). Against such importance he cites Marx, Dewey, and Don Ihde. Ihde, for example, is said to stress that "the practical perspective is...not only prior to but...more fundamental than the theoretical." In view of these alternative views, Limper is concerned that I may be "begging an important question concerning the role of purely theoretical or speculative reasoning in the development of modern technology."

Further, at the end of his paper, Limper returns to the theme of the possible weakness of ideas in historical change, attributing a "Whiggish" perspective to Whitehead and, implicitly, to me.

My reply is that whatever I am doing I am not *begging* the question of theoretical intelligence's role in modern technology and historical change. I am engaged in *wrestling with* the question, and my conclusion is not that of Marx. It is closer to Dewey than may at first appear. It is much closer to Ihde (and Limper?) than Limper seems to realize. Perhaps, finally, it is not identical. I do, indeed, try to give theoretical intelligence more of what I think is its due than some of Ihde's writings would suggest. But I join Ihde in declaring practical intelligence much the older and, in important ways, much the more fundamental aspect of mentality, especially when intelligence is appreciated as the complex, unified set of functions I have already affirmed. I may not in Ihde's sense be a full-fledged "materialist" in my view of technology, but I am certainly not in his sense an "idealist," either.

No more am I "Whiggish" in my view of the relationship between ideas and historical change. And neither was Whitehead, if his careful work in *Adventures of Ideas* can be added to the less developed remark cited from *Science and the Modern World*. I take seriously (as Whitehead did) the fact that ideas have consequences; but other realities shape history (as Whitehead also pointed out) even more obviously. That is, in Whitehead's shorthand, *both* the Barbarians and Christianity, *both* Steam and Democracy, shaped the course of events. There are coercive forces that push us (climate, weapons, plague, invention of means of production, natural resources) and there are attractive ideals that pull us (revulsion against slavery, the lure of general political enfranchisement, human rights, economic justice). To ignore the coercive forces, on the one hand, would be sentimental other-worldliness. It would be to forget about the causal order in which we find ourselves emmeshed. Every event exists within a matrix of objective inheritance, as is central to Whitehead's analysis of reality. But to ignore the attraction of ideals, or to explain ideas away as mere ideology, would be materialist reductionism, a partial view in violation (no less than sentimentalism) of the unforced reading of the facts of human history and individual careers. Events involving human beings – Whitehead would say all events – have their subjective aims as well as their objective inheritances. Neither aspect alone tells the whole story. This is the stuff of which politics, including the politics of technological societies, is made.

IV

Finally, a concluding word in reinforcement of Limper's challenges to future work. He is on target, I believe, concerning all three areas of work he calls for.

Apropos the immediately preceding point on countering "Whiggish" or reductionist views of history, Limper would like more explicit discussion of real-world politics. So would I! Developing the bipolar analysis proposed by Whitehead's paradigms of Steam and Democracy would provide a special way in which process philosophers may make a fruitful contribution.[1]

If we are to engage in political thinking that is grounded in adequate ethical theories, we need to heed Limper's second challenge: "to develop some better way to connect...broad metaphysical concepts with specific ethical principles..." (Limper, p. 33). That is, indeed, one of the most urgent tasks before philosophers today, especially those philosophers who take seriously the ethical issues of environment ("ecojustice") and long-term future goods and harms ("transgenerational justice") that are raised with particular poignancy in philosophy of technology. Here are indeed nodal points well worthy of our best thought.[2]

Finally, if we are to place our ethical theories in the context of the best worked out thinking we can do about reality as a whole – and ethical theory is irresponsible if it neglects the rock-bottom realities with which its prescriptions must deal – then Limper's first challenge to philosophers sympathetic to process thought: "to seek out and articulate contemporary models which support their metaphysical principles" (Limper, p. 32), is entirely to the point. Nothing, in the long run, is more important than finding our bearings, cognitive and valuational, within the welter of data surrounding us. This is an immensely difficult (and controversial) challenge, to keep our critical wits about us at the same time that we open our minds to the fullest degree of comprehensiveness. But well-chosen models can help organize coherent comprehension and guide effective action.[3] Therefore I must welcome Limper's insistence that we redouble our efforts and get on with the job.

University of Georgia

NOTES

[1] I have attempted to make some progress along these lines in an article, "Myths and Hope for Global Society," *Contemporary Philosophy,* forthcoming. This is a reworked version of an earlier article previously published as "Hope and Myth in a World of Scarcity," *Georgia Review* **33** (Fall 1978): 553–570. Much more needs to be done.

[2] I have attempted some work along these lines in "Personalism and the Dignity of Nature," *The Personalist Forum* **2** (Spring 1986): 1–28; also in "Moderation, Morals and Meat," *Inquiry* **29** (December 1986): 391–406; and in "Obstacles on the Path to Organismic Ethics: Some Second Thoughts," *Environmental Ethics,* forthcoming.

[3] I have attempted to address these matters in such places as "Mapping the Logic of Models in Science and Theology," *The Christian Scholar* **46** (Spring, 1963); "Metaphysical Error: Social Disorder," *The Review of Metaphysics* **31** (June, 1978); "Religious World Modelling and Postmodern Science," *Journal of Religion* **62** (July 1982); "Organizing Images and Scientific Ideals: Dual Sources of Contemporary Religious World Models," *Proceedings of the 8th International Wittgenstein Symposium* (August, 1983); and "Toward a Postmodern Science and Technology," in D. Griffin, ed., *Spirituality and Society: Postmodern Visions* (Albany, N.Y.: State University of New York Press, 1988).

PART II

DEFICIENCIES IN ENGINEERING ETHICS

TAFT H. BROOME, JR.

IMAGINATION FOR ENGINEERING ETHICISTS

1. INTRODUCTION

Has progress in the scholarship and teaching of engineering ethics begun to stagnate? This question, together with responses to it given by Steven Goldman[1] and Langdon Winner,[2] motivated this paper.

Goldman argues that the scholarship of engineering ethics can no longer prosper without a much broader understanding of the nature and scope of engineering than currently exists within the engineering ethics community. Winner observes that the use of engineering ethics case studies in typical courses for college students has little impact on students in getting them to think about basic issues. Implicit in these responses is the assumption that something new has to be created and injected into the field of engineering ethics to give it new life.

I will argue that the ideas needed to give engineering ethics new life have already been created. Having been created mostly by engineers, these new ideas are not accessible to or expressed in ways that capture the imaginations of philosophers. Thus, these ideas have generally not received sufficient attention to affect their injection into the scholarship and teaching of engineering ethics.

The purpose of this paper is to examine a set of ideas about the nature and scope of engineering that have not yet been exploited by philosophers. Ideas about engineering both as a scholarly discipline and as a profession are included. Focus is on such ideas as are judged to be imaginative in the sense that they are seen to possess potentials to provoke, excite, or enliven the imaginations of philosophers. The aim of this paper is to reveal these potentials to philosophers in ways that might enable them to inject new life into engineering ethics. I do not deny that engineers like myself can contribute to a revitalized engineering ethics; here I merely throw out ideas to tantalize philosophers.

Paul T. Durbin (ed.), Broad and Narrow Interpretations of Philosophy of Technology, 45–51.

2. CORPORATE TENURE

In his book, *Controlling Technology*,[3] Stephen Unger touches upon the idea of a tenure arrangement for engineers employed in corporations. His idea is to use the academic tenure arrangement provided college faculty as a model for a practical means of resolving the whistleblowing issue.

The whistleblowing issue is now an old one. It addresses the duty of engineers in situations where their employers' interests in profit are perceived to conflict with their own professional interests in Cicero's imperative to safeguard the public. In their efforts to obtain a theoretical resolution of this issue, engineers have stipulated, in one form or another, that, "He [the engineer] shall hold paramount the safety of the public." This is now in most professional societies' codes of ethics. Nevertheless, symptoms of unrest on this public paramountcy imperative abound.[4] And there remains considerable debate as to whether moral grounds for whistleblowing are sufficient practical grounds for getting even those engineers who subscribe to public paramountcy to do their duty in the face of possible retaliation from their employers. Unger's corporate tenure idea would seek to provide whistle-blowers reasonable protection from retaliatory firing and black-balling.

The corporate tenure idea has not been exploited by philosophers apparently because attacks upon it have not been resisted. For example, no response has been given to the claim that as academic tenure has visited the burden of "deadwood" upon college faculty, so too would corporate tenure promote laziness and incompetency among engineers and incite sensationalistic movements to gain for them press coverage and fame. Apparently no ethicist has considered whether this claim is supported by experiences with similar employment protections afforded engineering ombudsmen; nor have ethicists measured the degree to which the protections offered by corporate tenure are assured by the changing legal environment in America with respect to job security. Assuming that engineering whistleblowers have no protection from retaliation of their employers, could American businesses be compelled to set up tenure arrangements for their engineers?

Law does not provide that corporations can be forced into tenure arrangements with their employees. There have been no serious attempts to force tenure upon corporations in order to serve the interest of the public either by means of civil dissent or by exertion of any union-like pressures from engineering professional organizations. Moreover, according to John Ladd's defense of "corporativism,"[5] there is no moral basis for forcing tenure upon corporations.

We can, nonetheless, in order to make a point, explore two corporations. Imagine that they offer similar environmentally hazardous engineering services to the same consumer markets. Prices, quality, and other business factors are assumed equal. While corporation X has no tenure arrangement with its engineers, corporation Y has established one voluntarily. Both corporations are competing for a contract with a local government whose constituency is known to be both very sensitive to and well informed about environmental issues and generally distrusting of corporate guarantees of safety. An argument can be made that corporation X would be less competitive than Y in winning this contract. The point is that corporate tenure can be *imagined* as economically compatible with both the corporation's profit imperative and the engineer's safety imperative over the long run.

My point is this: any expectation of corporations of type X coexisting with competitors of type Y would reveal new potentials for philosophers who study whistleblowing. Engineers employed in type X corporations can be fired and blackballed, and thereby suffer extreme mental and financial stress as a result of blowing the whistle. This might support a utilitarian argument to justify their failure to blow the whistle in cases where the anticipated consequences fall short of death or severe injury to many of the affected parties. On the other hand, engineers employed in type Y corporations, having greater freedom than their competitors, could be more responsible – though they must also be aware of the consequences of allowing whistleblowing to run unchecked.

(Other new ideas about the nature of engineering and its distinction from science, which lend support to those who would oppose the Ciceronian imperative, will be discussed in some detail below. The new idea I want to mention there is the idea that the nature of engineering provides grounds for exposing the public to risks that exceed those associated with scientific experimentation. Where a literal interpretation of the Ciceronian imperative is bounded by scientific risk, the new interpretation would allow engineers to expose the public to greater risks.)

While the emphasis in type X corporations might be on encouraging engineers to obey a new or "soft" interpretation of the Ciceronian imperative, emphasis in type Y corporations would be on restraining them from carrying literal or "hard" interpretations of the Ciceronian imperative too far. The whistleblowing issue is a zero-sum game for type X corporations, but it could foreseeably be a win-win game for type Y corporations.

In sum, the postulate of a type Y corporation suggests new opportunities for philosophers to collaborate with business scholars and engineers for the purpose of formulating practical means of resolving the whistleblowing issue. What this engineer's thought experiment suggests is that philosophers involved in engineering ethics have not been imaginative enough; utilitarian considerations do not always work to the detriment of whistleblowing.

3. ENGINEERING AND ENVIRONMENTAL ETHICS

Engineers change the human environment: they carve the landscape to make way for highways, bridges, dams, and skyscrapers; exhausts from their automobiles, airplanes, and industrial smokestacks more than noticeably affect the atmosphere; and they deposit our municipal and industrial wastes into rivers, lakes, and oceans. Nevertheless, while a substantial body of scholarship on environmental ethics is accumulating in the literature, scarcely a handful of ideas from this scholarship has been injected into the knowledge base of engineering ethics.

Gunn and Vesilind[6] have contributed significant case studies to the environmental/engineering ethics knowledge base. Mario Salvadori[7] has considered whether engineers have a duty to science and thereby to mankind in cases where fossils, ancient pottery, or other artifacts possibly possessing scientific value are encountered in engineering excavations. Little has been done to exploit these ideas, or to do more in connecting environmental ethics to engineering. Therefore, a vast set of opportunities apparently exists for philosophers to explore new territories.

One opportunity is presented by the exploration of space. Surely cosmic space is part of the human environment. Pollution of space is becoming a concern in response to suggestions for using it as a nuclear waste depository and as a high-tech battleground. A collection of member nations of the United Nations put forward their Common Heritage of Mankind principle[8] which suggests that space belongs to everyone – to exploring nations as well as to non-exploring nations – and thus the "spoils" from space must be shared among all peoples. Although the U.S. does not endorse this principle, it can be viewed as an extension of the law of the sea. It poses a problem for environmental ethics by asserting that space can be owned, or that its use by some people must respect the interests of all people. The challenge space presents to philosophers is for them to examine the moral dimensions of using space *before* its use becomes unacceptably hazardous. Clearly, a problem of distributive justice is posed. Again this is a largely speculative possibility, but even a cursory perusal of the engineering ethics literature will demonstrate that it has been ignored – though not entirely ignored by other philosophers.[9]

Another example. The Gaia hypothesis affords philosophers the opportunity to engage renewed discussions on the subjects of wholism and organicism. Though controversial, the Gaia hypothesis is firmly rooted in the physical sciences, and it suggests that the earth's biota are major participants in the earth's climate control mechanism. Moreover, the hypothesis suggests that the biotic system tends to regulate climatic conditions in ways favorable to itself. In view of the rate at which the hypothesis is gaining support within segments of the scientific community, a new wholistic or organic starting point for ethical inquiry into human relationships with the environment can be anticipated. Since engineers are "experts" on changing the environment, the Gaia hypothesis could become a useful starting point for new investigations among philosophers into old environmental issues. For example, while environmentalists are poised to appeal to the Gaia idea for support of their claim that polluting is bad, polluting corporations are prepared to argue that polluting is not bad on appeal to the Gaia idea for support of their claim that the biota will compensate for the ill effects of pollution. Philosophical problems tend to be open-ended, and this one surely is.

4. NEW IDEAS ABOUT THE NATURE OF ENGINEERING

Until recently, virtually every use of the term "engineering" has referred to something like "the application of science to practical or societal problems."[10] The Kotarbinski school of praxiology defines engineering as the science of efficient action,[11] and other views likewise declare engineering to be one or another sort of science.[12] Together with others, I have argued that engineering cannot be an applied science nor any other kind of science,[13] but a member of a separate learned discipline having kinships with medicine, law, and politics.

This new "non-science" view of engineering is supported by the Krohn and Weingart[14] argument that engineering systems are typically so complex that knowledge about their behaviors can be obtained only from tests activating the full range of the systems' operating conditions. Martin and Schinzinger[15] have argued, therefore, that engineering systems are experiments with the public as human subjects. Probably because the affected parties of engineering experimentation are more difficult to identify than the parties normally affected by scientific or medical experimentation, philosophers have missed an opportunity. They might, for instance, have been among the first to observe that the astronauts were clearly the affected parties in the shuttle Challenger disaster; therefore they were the persons who should have been notified about the 0-ring hazards.[16]

Henry Petroski[17] argues that engineering design advances primarily as a result of engineering failure. Accordingly, the new view of engineering would assert that progress in space technologies can not be possible without the implementation – and failure – of complex engineering systems subjected to full operating conditions in the space environment. Another example: the Strategic Defense Initiative (SDI) could be judged feasible from this new engineering standpoint, while at the same time it might be judged infeasible from the scientific point of view, that knowledge about SDI can not be obtained prior to its implementation.

Contrasts of these two points of view could be used as starting points for renewed debates on moral issues related to technologies that have enormous implications for human existence. Philosophers may be well advised to consider expansions of these debates: e.g., moral implications of superconductors, networked supercomputers (and their viruses), manufacturing in space, and other technologies anticipated for the year 2000.

5. CLOSURE

The starting point for this paper is the view expressed by some that progress in the scholarship and teaching of engineering ethics has stagnated. The main thesis of this paper is that the ideas needed to give engineering ethics new life have already been created, but they require the attention of philosophers before gaining entry into the scholarly literature. The aim of the paper, therefore, was simply to address some of these ideas in ways that might stimulate the imaginations of philosophers, trying to invite them to exploit the potentials of these ideas to generate growth in the field of engineering ethics.

The significance of this paper, finally, lies not only in its effort to increase productivity in engineering ethics. It is also an attempt to illustrate the need to breed new scholars equally able to appreciate imaginative ideas and to inject them into an understanding of the nature of engineering that will revitalize the engineering ethics literature.

Howard University

NOTES

[1] Steven L. Goldman, "Philosophy, Engineering, and Western Culture," this volume, pp. 125–152; an earlier version was presented by Goldman at a symposium from which the present set of papers emerged.
[2] Langdon Winner, "Engineering Ethics and Political Imagination," this volume, pp. 53–64.
[3] Stephen Unger, *Controlling Technology: Ethics and the Responsible Engineer* (New York: Holt, Rinehart, and Winston, 1982), p. 60.
[4] Taft Broome, "The Slippery Ethics of Engineering," *The Washington Post*, Sunday, December 28, 1986, p. D3.
[5] John Ladd, "Corporate Mythology and Individual Responsibility," *International Journal of Applied Philosophy*, Spring, 1984, pp. 21–22.
[6] A. Gunn and P.A. Vesilind, *Environmental Ethics for Engineers* (Chelsea, Mich.: Lewis, 1986).
[7] Mario Salvadori, "Historical Sites and the Construction Engineer," *Journal of Construction* (ASCE) **102** (June 1976): 295–301.
[8] H.W. Cannon, *Agreement Governing the Activities of States on the Moon and Other Celestial Bodies* (Part 4), Committee on Commerce, Science, and Transportation, U.S. Senate, November, 1980, pp. 452–454.
[9] See, for example, the articles under the heading "Philosophical Problems of Space Exploration," *The Monist* **71** (October, 1987).
[10] Peter Caws, "Praxis and Techne," in G. Bugliarello and D. Doner, eds., *The History and Philosophy of Technology* (New York: Harper and Row, 1978), pp. 227–237.
[11] Tadeusz Kotarbinski, *Praxiology* (Oxford: Pergamon, 1965).
[12] See, for example, Ian Jarvie, "Technology and the Structure of Knowledge," in C. Mitcham and R. Mackey, eds., *Philosophy and Technology* (New York: Free Press, 1972), pp. 54–61.
[13] Taft Broome, "Engineering the Philosophy of Science," *Metaphilosophy* **16** (1985): 47–57.
[14] W. Krohn and P. Weingart, "Commentary: Nuclear Power as a Social Experiment – European Political 'Fall Out' from the Chernobyl Meltdown," *Science, Technology, and Human Values* **12** (Spring 1987): 52–58.
[15] Mike Martin and Roland Schinzinger, *Ethics in Engineering* (New York: McGraw-Hill, 1983).
[16] S. Goldberg, "The Space Shuttle Tragedy and the Ethics of Engineering," *Jurimetrics Journal* (Winter 1987): 155–159. See also Broome, "The Slippery Ethics of Engineering" (note 4, above), for the same conclusion arrived at independently.
[17] Henry Petroski, *To Engineer Is Human: The Role of Failure in Successful Design* (New York: St. Martin's Press, 1985).

LANGDON WINNER

ENGINEERING ETHICS AND POLITICAL IMAGINATION

Recent attempts by American colleges and universities to teach ethics for scientists and engineers deserve strong praise. They represent a shift away from the idea that questions about ethics and morality are best left to humanists or to elder statesmen of science, a recognition that such matters ought to be an important part of education in the technical professions. One can hope that through these efforts a new generation of men and women will obtain a firm grounding in the ethical aspects of their vocations early enough to make a difference.

Despite these admirable aims, however, the approach often used to teach ethics to scientists and engineers leaves much to be desired. In the way the topic is usually presented, personal responsibilities are situated in extremely limited contexts and ethical choices made to seem something like extraordinary, unwelcome intrusions within a person's normal working life. Rather than lead students to evaluate the most basic, most practical features of their career choices – the kinds of work they select and the social conditions in which that work is done – courses on professional ethics tend to focus upon relatively rare, narrowly bounded crises portrayed against an otherwise happy background of business as usual.[1]

One way in which college courses avoid the difficult underlying questions that technology oriented professions involve is to focus upon case studies of particular ethical dilemmas. This can seem to be nothing more than a useful attempt to transcend mere abstractions and to provide contexts for issues by locating them in the "real world" of practice.[2] Unfortunately, what such moves often do is to bracket the realities of daily work in favor of hypothetical situations that are comforting because they are so remote. Cartoons of concreteness become the abstractions at hand. Hence the tone of such pedagogical case studies is often something like the following.

"You are an engineer working for a defense contractor helping to assemble the latest version of the cruise missile. One day you discover that the paint used on the shell of the missile is emitting toxic fumes that may be dangerous to people working in the assembly plant. The project is behind

Paul T. Durbin (ed.), Broad and Narrow Interpretations of Philosophy of Technology, 53–64.

schedule and your boss has made it clear that it must be completed within deadline. Should you blow the whistle on the toxic fumes or keep silent thereby avoiding risk to your job and career?"

I have exaggerated, but not by very much. So-called ethics case studies usually point students toward specific troubling incidents within what are assumed to be otherwise harmonious patterns in ongoing institutions. The patterns themselves, however, are not identified as anything problematic. Indeed, it is a property of the case study approach to education in business, law, and engineering that the contexts that underlie particular cases are never themselves called into question. By failing to analyze and criticize these contexts, case studies tend to legitimate and reinforce the *status quo*. Thus, the decision to work for a defense contractor building thermonuclear weapons can well go unnoticed as an ethical issue at all. A student may learn how to deal with the moral dilemmas, perhaps even ones that involve great personal courage. But missing from the inquiry are some important dimensions of the lives of technical professionals.

POWER AND THE "GOOD SLOT"

A story from my own experience with engineering students will help illustrate the point I am making. While it is only one incident, it reflects situations I have encountered all too often.

Several years ago I taught a course on the history of technology, co-teaching it with a professor of aeronautical engineering. We had reached the part of the term in which we studied the founding of the modern engineering professions and engineering education. That history is, of course, one in which the engineering schools and large corporations work closely together to tailor an engineering curriculum suited to the immediate needs of the corporations.[3] After the class had gone over the facts in the readings, my colleague turned to the group of about thirty science and engineering undergraduates and asked a question that I would probably never have had the nerve to raise.

"Well," he said, "how does it feel to be on a conveyor belt being turned out into slots?"

There was a long embarrassing pause. Finally, one of the students, a fellow who had been very quiet most of the term, raised his hand and answered.

"I don't want to be put into any slot," he said confidently. Then he thought for a moment and added, "I want to find a good slot!"

The classroom filled with laughter; seldom is a point so eloquently made.

But the story does not end there. During the ensuing months I got to know this fellow fairly well. We took long walks by the Charles River and talked about his life and the choices he faced. It turned out that he was, in a pattern fairly typical of engineering undergraduates, the son of a machinist and had gone to college to fulfill his working class family's dream of upward social mobility. He talked about his major, electrical engineering, and expressed his qualms about some of the uses being made of computers these days and of his desire to use his skills for something socially beneficial.

Since I always encourage students to think about the social and ethical dimensions of their work, I thought to myself: "Good. Here is someone I have really reached."

About a year later during the summer break I ran into the fellow walking across campus and asked how he was doing.

"Very well," he replied, "I've got a great job."

"Where is it," I inquired.

"The Draper Lab," he said.

"Oh. What do they have you doing?" I asked.

"Computer programming," he said and went on to explain that it was programming of an especially intricate, challenging kind.

"And what are the programs used for?" I probed.

"Guidance systems," he replied.

My next question was obvious. "And what do these guidance systems guide?"

"Intercontinental ballistic missiles," he allowed with an uncomfortable smile spreading across his face.

"Bruce, you're building bombs, aren't you? Tell me, on the basis of your own personal commitments, how you can justify that."

He went on to list the kind of deals today's small time Fausts tend to make with today's big time Mephistopheles: the problems were challenging in a purely technical sense, "cutting edge" stuff. The recognition by senior colleagues was gratifying, especially for someone who had never had much recognition. And, of course, you just could not beat the money they were paying.

I told him of my disappointment at what he was doing and silently pondered my contempt for a system that would lure a fine young fellow like himself into the company of those who build the instruments of war and death.

"I guess you've found your *good slot,*" I groaned and we went our separate ways.

There is, in my view, a tendency in our engineering professions as presently constituted not to see the kind of work they do as involved with serious moral questions. Convinced that technology is merely neutral, persuaded that somehow the important questions will be raised by somebody else, many engineers seem almost proud of their inability or unwillingness to talk about ethical concerns. They are, by and large, more than ready to become the passive tools of whichever corporation or bureaucracy makes the highest bid.

Students do not enter our colleges saying to themselves: "Oh how I wish I could learn to build hydrogen bombs. If only I could learn the techniques of producing environmental carcinogens. Teach me the methods that will help an automobile firm decide whether or not it is cost-effective to replace gasoline tanks that explode on impact." No, they do not enter engineering school affirming goals like these. But what do they end up doing when they leave and look for a job?

The condition I am describing is not, I want to emphasize, one that characterizes engineering alone. In other prominent professions of our age we see similar circumstances in which people no longer see themselves as self-conscious moral agents and citizens of a free society, but rather as subjects of large corporate bodies that, while extending organizational control, gradually eliminate the individual's need to think.

From this point of view, one element notably missing from the education of technical professionals is any critical encounter with a person's relationship to power. Indeed, the phenomenon of power and one's involvement in it is perhaps the best kept secret in modern education generally. Courses and whole curricula simply assume that certain people and institutions hold power; such people have the jobs and define the possibilities for a working life in engineering. Hence a lecture or problem set in a science or engineering course may begin with the words: "Assume you are working for a defense contractor." "Assume you are employed by a large corporation." Such contextualizing statements assume that a student need not ask such prior questions as: Is this the way it ought to be? Is that the proper setting for the practice of my hard-earned knowledge and skill? How might it be otherwise?

The tacit premise of the case study approach to professional ethics is that as one enters a profession, one simply embraces the existing commitments, institutional patterns, and power relationships the profession contains. There is no understanding that as one ponders joining the enterprise, one also has an opportunity to re-evaluate or even reconstitute that field. Only after the newcomer's basic compliance has been secured is he or she assumed ready for the bracing ethical challenges that lie ahead.

2. A SENSE OF VOCATION

The failure of engineering ethics courses to explore the phenomenon of power reveals another significant lack: exploration of the question of vocation, one's calling in a moral sense.[4] It is reasonable to expect that as a person contemplates committing several decades to a profession, some basic issues ought to be addressed. What are the fundamental ends of a life invested in this line of work? What is the purpose of developing my knowledge and skill in this direction in the first place? Who ought to control the most basic definitions of what my vocation entails? Our educational institutions now shortchange students by avoiding such issues, neglecting focused study of the moral and political groundwork of professional life.

Courses on engineering ethics tend to focus upon issues of right and wrong in personal conduct – extremely important matters indeed. But beginning with Aristotle, philosophers have noticed that there is a logical juncture where ethics finds its limits and politics begins. That turning point comes when we move beyond questions of individual conduct to consider the nature of human collectivities and our membership in them. This calls upon us to ponder the nature of political society and what membership in it means for us, not merely as individual actors but as participants in a community.

In my own teaching of engineers and scientists I find some who are actually prepared to act heroically at the level of individual ethics, but who have no sense whatsoever what their work means in an organizational or political sense. It comes as a surprise, for example, for students to learn that engineering projects tend to reflect some economic or social interests and not others. They learn next to nothing about the history of their fields, about who works for whom and how that came to be. These are, once again, well kept secrets. By the time one gets to be a graduating senior, all he or she may want to know is: Where are those nice people with all the high paying jobs?

The lack of an ability among engineers and other technical professionals to engage in political reflection is by now much more than a small failure in education; it threatens to become a great tragedy. We live in an age of scientific technology. Many of the most significant ways in which old forms of power are reproduced and new ones created are reflected in the technologies we use. To a great extent the possibilities of social and political life in the twentieth century are defined by technological opportunities and constraints. Whether we are to have a good society or a bad one is powerfully influenced by the technologies we develop and put to use. For that reason the role of engineers and technical professionals is crucial. They are intimately involved in maintaining key social patterns and in inventing new ones as well. In that work they become, in effect, unelected delegates and representatives of the rest of us, charged with the work of building basic structures of our social and political future. If it turns out that they lack vision, lack the ability to make choices that express not only knowledge but also personal strength born of wisdom, our society is headed for trouble.

From this vantage point what is needed is less the study of pre-packaged ethical dilemmas than the cultivation of two crucial skills: political savvy and the capacity for political imagination. At some point in their education engineering students need to examine critically the historical origins of their own chosen profession. They need to understand how their branch of engineering was first organized, by whom, and for what reason. They need to examine the kinds of social and economic interests their field represented at the beginning as well as the ones it expresses now. And they need to understand who controls the choice of projects and why.

Casual observers might suppose that learning to become savvy about the social realities of the world students are entering would be a standard part of a good engineering education. In fact, it is almost never explicitly addressed. What students receive instead is a kind of hidden curriculum, a set of unstated assumptions, expectations, and role models which provide a subtle but thorough enculturation. Its underlying text is: Do not ask questions; just find your place in the established hierarchy and obey orders.

There are many ways in which political savvy of this kind might be nurtured and even tested. In my own courses, for example, I have sometimes used the following quiz.

Read carefully Lewis Mumford's chapter, "The Nucleation of Power," in *The Pentagon of Power*[5] and answer the following questions:

1. Do you consider Mumford's description valid or not?
2. Will you in all likelihood join "The Power Complex"?
3. Explain why.

Of course, achieving political savvy alone is not sufficient; it may leave a person aware but cynical, thus accomplishing more harm than good. For that reason there is a need to nurture more – political imagination, the ability to envision the contributions of one's work to society as a whole, to the quality of public life. Within the plans, methods, and prototypes an engineer produces are the blueprints of our social future. This is true whether these products have to do with water projects, highways, industrial machines, electronic components, or energy systems. As part of mastering the fundamentals in their fields, engineers and other technical professionals ought to be encouraged to ask: Can we imagine technologies that enhance democratic participation and social equality? Can we innovate in ways that help enlarge human freedom rather than curtail it? How can planning for technological change include a concern for the public good as distinct from narrowly defined economic interests?

An ability to investigate questions like these ought to be part of the intellectual tool chest that engineers carry with them when they leave the university. While there is no established curriculum that imparts such skill, many of the disciplines in the humanities and social sciences have important resources to offer, resources that at present remain largely untapped. In political philosophy, for example, it is possible to build bridges to the study of engineering by showing ways in which political philosophers are to some extent designers. From Plato to Jean-Jacques Rousseau to John Rawls, political theory has involved an attempt to translate the ideals of the good society into workable institutional structures. Those who met at the Philadelphia convention of 1787, for example, sought to invent a political machine, the U.S. Constitution, that would provide beneficial results generation after generation. To some extent, therefore, political theory and engineering share a common project – making things that will endure. It might even be argued that they share some of the same criteria of evaluation: a concern for the appropriate furnishings of the good life.[6]

Whatever specific fields in the humanities may be chosen to illuminate such questions, however, it is crucially important to allow our students to become competent not only as scientific and technical problem-solvers (the only need currently recognized by the academy and business), but as dreamers and visionaries as well.[7] Drawing upon materials in their own fields, they should be encouraged to imagine the best possible uses to which their intelligence and know-how might be put. Engineers and technical professionals are the unacknowledged legislators of our technological age. Choices that affirm the public good or trample it often rest in their hands. If they overlook this critical role and responsibility, they will also acquiesce in yielding power to agents whose ends are increasingly distant from humanity's best.

3. OURS IS NOT TO REASON WHY?

Much of the vocation of engineering is legitimately, admirably involved with questions of "how?" – how to solve problems, how to achieve particular results within physical and economic constraints, how to get things done. Such concerns are and will always be central to an engineer's daily activity. Providing students the wherewithal to grapple effectively with "how" questions is bound to remain the heart of engineering education. But in a world of rapidly advancing technology, issues about "how" are often less important than ones that concern "why?" Our society has increasingly powerful means at its disposal and a great many expediencies that seem to justify applying these means all over the planet as rapidly as possible. But typically missing from our deliberations about new technical means are deeply grounded reasons to guide our choices.

Yes, it is possible to build the next intriguing generations of space-based offensive and so-called defensive weapons. In the face of pressing human needs in other domains, why invest our scarce resources in things designed only to destroy? Before going further, it is crucial that we examine the underlying rationale that governs policy. Perhaps that pause for reflection would suggest alternatives other than blindly continuing the arms race.

Yes, it now appears possible to renovate the genetic structure of life forms on the planet. But why? In what sense are such projects needed? What ends or purposes are to be served? Before going ahead, one needs to explore the reasons.

And yes, it is possible to build any number of nearly or totally automated systems of production. But why? Which ends will be furthered? Who stands to benefit and who to lose? Before we engineer systems of this kind, we need to be clear about the grounds that justify our projects.

As one examines such topics, the question of "why" can be reformulated in a number of useful, challenging ways. One can ask: For the sake of what deeply meaningful ends are our technologies well suited or ill suited? In response to what central human needs are our techniques and instruments developed and applied? On the basis of what fundamental orienting principles do our technology based practices and institutions find their uses and their limits? In selecting a particular engineering project, what kind of world do we affirm and seek to create?

There are bound to be strong disagreements in our answers to any of these questions. Indeed, it is no easy matter to set about clarifying the basic notions that ought to guide the development and use of our technical means. People have widely different understandings about the meaning and proper application of terms like freedom, justice, security, human rights, well-being, public good, and other key concepts. But it is crucial that such ideas be continually discussed and debated throughout our deliberations about key technological choices. If we do not do this, ideas about "how" begin to comprise the whole of our thinking. Indeed, as Jacques Ellul has argued, it appears to be the destiny of modern thought to replace any living concern about human ends with sterile discussion about instrumentality and efficiency.[8]

A stunning example of this tendency was the Star Wars debate of the early 1980s. President Reagan announced the Strategic Defense Initiative in March, 1983, proclaiming its noble goal for the United States, that of "rendering nuclear weapons impotent and obsolete." From that moment forward most of the debate, for and against, focused on the question: Will it work? Will the technologies function as planned? In the heated debates that followed, people seized upon the instrumental issues – issues about "how" – as if they were the truly essential ones.[9] Looking back on this period of history, it is interesting that S.D.I. was offered at a time in which there was, for a while, intense public questioning around the world of the very ends of defense policy, questioning expressed in the Nuclear Freeze movement and in widespread discussion of Jonathan Schell's book, *The Fate of the Earth.*[10] Reagan's move was a clever one. Once again it got people back to talking about gadgetry and instrumental concerns and far away from thinking about fundamental purposes and the relationship between ends and technical means.[11]

This unwillingness to explore the basic reasoning behind our policies accounts for much of the absurdity that passes for informed discourse about technological choices nowadays. Thus, there is much attention to "cost/benefit" analysis, but without the slightest regard for the ultimate foundations of our judgments of benefit and cost. Similarly, one finds people nodding in confident agreement about "efficiency" without pausing to consider what is being maximized and for what reason; that is, what defines our numerators and denominators and on what grounds? Or one hears a great deal of praise for "progress" which seems to assume that one is progressing along a particular path, but never is the path or its destination specified. Similarly, there are anguished cries for increased national "competitiveness" with no attempt to justify the desirability of the things at stake in the competition.[12]

In all discussions of ambitious proposals for technological change, inevitably the most revealing and embarrassing question one can ask is simply, "Why?" What one discovers in a great many cases is that the answers have somehow been ignored, forgotten or even suppressed. That is why any program of engineering education that does not equip students insistently to inquire into the basic purposes of their work is simply an affront to reason. Any technical analysis that asks "how?" without first considering "why?" is bound to be vacuous.

With some ingenuity, the "why" could well be included in difficult technical courses, not just humanities electives set aside in a separate, easily forgotten corner of the curriculum. For example, one useful strategy in doing cost/benefit analysis is to begin by listing costs and benefits and, before plowing on, simply reverse the labels over the two columns to see how things look. Analyses by transportation engineers during the 1950s counted "increased traffic flow" as a benefit, something that communities affected by highway planning would later count as a significant cost. To approach cost/benefit analysis by reversing labels always causes a great commotion and you seldom get asked back, but it can add some life to the otherwise rigid, routinized procesess of technocratic thinking.

4. ENGINEER AS CITIZEN

In sum, the great need at present is for our professionals to take a mindful, critical stance with respect to their own work. This requires, first of all, that a person scrutinize the projects in his/her field and the consequences of those projects within the context of a continuing dialogue about our culture's most cherished ideals and principles. It implies, secondly, that scientific and technical professionals must locate and, when necessary, actively create public roles in which one's understanding of important conditions can become a focus of new thinking, debate, and action for the public as a whole. Ethical responsibility now involves more than leading a decent, honest, truthful life, as important as such lives certainly remain. And it involves something much more than making wise choices when such choices suddenly, unexpectedly present themselves. Our moral obligations must now include a willingness to engage others in the difficult work of defining what the crucial choices are that confront technological society and how intelligently to confront them.

The first requirement, therefore, might be called the responsibility of dialogue; the second, the responsibility of citizenship. Both of these were, of course, well understood by the ancient Greeks. Our task is that of bringing these practices to life for the era of high technology. It is obvious that in fact these are not responsibilities that are widely cultivated or practiced. At present, engineering students are encouraged to become careerist and self-interested in ways that exclude dialogue altogether. Their university degrees could well carry not only the title "Bachelor of Science," but also "Certified Uncommitted."

Nevertheless, considerable hope can be found in a single fact: it does not require many people to turn things around. Even one voice speaking at the right moment and in the right area can make all the difference. The contribution of Rachel Carson in the early 1960s was an instance of this kind.[13] At first widely denounced by much of the community of scientists and chemical engineers, Carson's message foreshadowed an international environmental awakening and a gradual move away from the use of destructive chemicals in agriculture.[14] A similar sense of ethical commitment was evident in the petition drive by many scientists and engineers of the mid-1980s, pledging not to work on any part of Reagan's Strategic Defense Initiative. Organizations such as Physicians for Social Responsibility, Computer Professionals for Social Responsibility, the Union of Concerned Scientists, and other groups have set out to raise public consciousness about important public policy issues in their fields. A number of computer scientists in the United States and Europe, for example, have begun devoting their attention to ways in which new information systems might protect freedom rather than undermine it.[15]

Examples like these suggest what I understand to be the genuine promise of the exercise of responsibility among scientific and technical professionals: a gradual reorientation of patterns of research, development, and application of emerging and already existent technologies to accord with our civilization's higher principles. It is "progress" conceived in this manner, rather than melodramatic rehearsals for whistle blowing, that offers the real challenge for engineering ethics in our time.

Those who acknowledge the two kinds of responsibility I have mentioned, those of dialogue and citizenship, face an unsettling paradox. On the one hand it is clear that, properly speaking, a person can be responsible only for his or her own decisions, actions, and their consequences. At the same time there is an important sense in which each person is now responsible for nothing less than the future of humanity itself. Confronted with that paradox and the burdens it carries, many people are inclined to flee into the security of their private lives and the satisfactions of narrowly defined competence, areas where they feel they have at least some safety and control. That impulse, more than any other cause, is what renders technical professionals morally impotent, ready to be manipulated by whatever enticements the state and large corporations have at their disposal. Any effort to define and teach engineering ethics which does not produce a vital, practical, and continuing involvement in public life must be counted not just as a failure, but a betrayal as well.

Rensselaer Polytechnic Institute

NOTES

[1] See, for example, Stephen H. Unger, *Controlling Technology: Ethics and the Responsible Engineer* (New York: Holt, Rinehart and Winston, 1982).
[2] For a defense of this method see C. Roland Christensen *et al., Teaching and the Case Method* (Boston: Harvard Business School Press, 1987).
[3] David F. Noble, *America by Design: Science, Technology and the Rise of Corporate Capitalism* (New York: Knopf, 1977).
[4] Max Weber explores a similar issue in "Science as a Vocation," in H.H. Gerth and C. Wright Mills, eds., *From Max Weber* (New York: Oxford University Press, 1946).
[5] Lewis Mumford, *The Myth of the Machine: The Pentagon of Power* (New York: Harcourt Brace Jovanovich, 1970), chapter 9.
[6] I discuss this topic in "Techne and Politeia," in *The Whale and the Reactor* (Chicago: University of Chicago Press, 1986), chapter 3.
[7] An early but overly optimistic attempt to portray computer scientists in this light can be found in Robert Boguslaw, *The New Utopians: A Study of Systems Design and Social Change* (Englewood Cliffs, N.J.: Prentice-Hall, 1965).
[8] Jacques Ellul, *The Technological Society* (New York: Knopf, 1964).
[9] An exception is John Tirman, ed., *Empty Promise: The Growing Case against Star Wars* (Boston: Beacon Press, 1986).
[10] Jonathan Schell, *The Fate of the Earth* (New York: Knopf, 1982).
[11] An excellent discussion of Reagan's rhetoric here is Janice Hocker Rushing, "Ronald Reagan's 'Star Wars' Address: Mythic Containment of Technical Reasoning," *Quarterly Journal of Speech* **72** (1986): 415–433.
[12] See Martin Kenneth Starr, ed., *Global Competitiveness: Getting the U.S. Back on Track* (New York: Norton, 1988).
[13] Rachel Carson, *Silent Spring* (Boston: Houghton Mifflin, 1962).
[14] Keith Schneider, "Science Academy Says Chemicals Do Not Neccesarily Increase Crops," *New York Times*, September 8, 1989, p. 1.
[15] I discuss such attempts in "Political Ergonomics: Technological Design and the Quality of Public Life," *Wissenschaftszentrum Discussion Papers*, IIUG dp 87–7 (Berlin: Wissenschaftszentrum Berlin, 1987).

PART III

SYSTEMS THEORIES

SYTSE STRIJBOS

COMPUTER AND WORLD PICTURE: A CRITICAL APPRAISAL OF HERBERT A. SIMON[1]

This article is related to a broader study of systems thinking[2] which investigates how this modern, influential movement of thought is related to the technical world picture that has dominated Western thought since René Descartes. The present article limits itself to a critical appraisal of Herbert A. Simon as a systems thinker and how his scientific work evinces a view of reality inspired by computer technology.

Two main aspects are distinguishable in Simon's work. First, in creating his particular field of specialization, Simon developed a systems theoretical framework. Fundamental to this is a distinction between two spheres of reality: natural and artificial. The term "artificial" is commonly used in connection with technology. Simon, however, gives it a much broader meaning. As Simon sees it, the artificial world includes the whole of human culture. This entire realm can be constructed in a technical way. Even the process of human thought is so interpreted: human thought is narrowed down, in a reductionist, technicist manner, to the behavior of a computer, an artificial (in this case information-processing) system.

Such an equalization of human and computer is fraught with far-reaching consequences. Reflection on these consequences forms the second main aspect of Simon's work. He has devoted special attention to the effects of computerizing organizations. His reductionist, technicist view of the human and of interhuman relations, I will argue, deprives him of a satisfactory perspective on authority, freedom, and responsibility within organizations.

1. PRELIMINARY OBSERVATIONS

There was never a time or a culture in which technology occupied so central a place as it does in Western society today.[3] Nor is this technical culture any longer restricted to the Western world. Seemingly without effort it appears to be growing, unlike any other in human history, into a global culture overarching a diversity of societies, traditions, and religions.[4] However impressive and overwhelming the power and possibilities of modern technology may be,

Paul T. Durbin (ed.), Broad and Narrow Interpretations of Philosophy of Technology, 67–86.

the rise of this global, technical culture cannot be accounted for from that alone. Technology owes its central position primarily to the spiritual background of our culture, *viz.* to a technical way of regarding reality that has dominated Western thinking for centuries. To gain a clearly focused view of modern times one therefore does well not to pay exclusive attention to technology and its problems but to look as well at the *technical world picture*.

The history of the technical world picture, which has often been referred to by the narrower term "mechanistic," is closely related to the rise and development of classical natural science. Therefore this world picture, which E.J. Dijksterhuis discusses in his famous historical study, *The Mechanization of the World Picture*,[5] can also be described as a scientific world picture. Various scholars have investigated how the technical or scientific world picture evolved in association with the progressive development of natural science and technology. Thus in the seventeenth century models were derived from the construction and operation of a mechanical clockwork and in the nineteenth century from the steam engine. Still rather recently – since the fifties in our own century – the rise of systems thinking and computer technology has produced another drastic alteration of the world picture.

Fundamental to the claim of Ludwig von Bertalanffy, the father of systems thinking, and a group of thinkers associated with him (Kenneth Boulding, R.L. Ackoff, Ervin Laszlo) is the idea that systems thinking overthrows the centuries-old technical world picture, puts a new world picture in its place, and so offers a new perspective in a culture scarred by scientization and technicization. The "organismic trend" in systems thinking, upon which von Bertalanffy pinned his hopes, can, however, be criticized as a disguised form of technical reductionism.[6]

The work of the systems thinker and computer expert Herbert A. Simon occupies a somewhat distinctive place in the systems field. Simon cannot be included in the group of systems thinkers associated with von Bertalanffy, who are more or less aware of the dangers inherent in the technical world picture and intent on breaking its spell. On the contrary, Simon may be characterized as one of the most fervent and also extreme propagandists of the new technical or scientific world picture inspired by computer technology and systems thinking. Examination of his views indicates how far the main stream of systems thinking since von Bertalanffy has diverged from his deepest, original intentions. Not only is the organismic trend he advocated no different in principle from the mechanistic trend he opposed, it has also proven to be of little importance.

2. HERBERT A. SIMON

The broad research interests of Herbert A. Simon, who was born in Milwaukee, Wisconsin, in 1916, span the disciplines of computer science, psychology, and business adminstration, as well as economics. This professor at Carnegie-Mellon University, Pittsburgh, thus seems to be the ideal interdisciplinary systems thinker. He began his scientific career with a now classic study entitled *Administrative Behavior* (1947),[7] which deals with decision-making processes in economic organizations. For his pioneering work in this field he was awarded the Nobel Prize in economics in 1978.

Central in Simon's approach is the role played by rationality in human acting and deciding. In opposition to the prevalent view in economic theory that decisions must be regarded as the result of *perfect, unbounded* rationality, Simon developed the view that human rationality is *restricted and bounded.* Indeed, he considers the theme of rationality, in particular of the possibilites and the limits of human reason, as the heart of all his scientific work. Looking back almost forty years later, Simon opens *Reason in Human Affairs* (1983) with the statement: "The nature of human reason – its mechanism, its effects, and its consequences for the human condition – has been my central preoccupation for nearly fifty years" (p. vii).[8]

A few years before the appearance of *Administrative Behavior,* the first electronic digital computer was built. In the years after 1947 modern systems approaches – cybernetics, general systems theory, and the like – made their appearance in the sciences. Both systems theory and computer technology were of great importance for Simon's research activities from the fifties onward.

It is important to notice that Simon was intimately involved with the introduction of the computer into modern organizations. At first the function of the computer involved little more than replacing existing bookkeeping and key punch machines for processing administrative data in the organization. Gradually, the task of the computer expanded to support decision making at various levels in the organization. By applying such mathematical methods as operations research, it proved possible to formalize particular human functions so that they could be automated. A trivial example of a routine decision by the computer is printing a standard letter in cases of nonpayment of bills. Possibilities extend far beyond this, however. The short- and long-term prospects were first presented by Simon in his book, *The New Science of Management Decision* (1960).

One can see that the new technological possibilities of the computer provided the occasion for opening up new fields of scientific research. The application of the computer in decision-making processes confronts the researcher with fundamental questions, such as: How do people think? How do they solve problems? Linked directly to such questions is another: To what extent can human thought processes be transferred to the computer? In the fities, Simon and Allen Newell carried out pioneering scientific studies bearing on these questions. Their collaboration resulted in fundamental contributions to cognitive psychology and to the field known today as artificial intelligence.

The basic idea underlying Simon and Newell's investigations is that both humans and the computer can be regarded as information processing artificial systems. An elaboration of this basic idea is presented in *The Sciences of the Artificial* (1969, revised edition 1982). Here Simon aspires to advance a general theory of artificial systems. As he says of this work later: "I described properties that are common to all adaptive (artificial) systems, giving us a basis for constructing *a general theory of such systems" (Reason in Human Affairs,* p. 5, italics added).

These points are dealt with more fully below. Simon's systems-theoretical framework of thought is examined first, by way of analysis of the fundamental distinction between the natural world and the artificial world and Simon's view of the sciences of the artificial. Following this, Simon's view of the implications of the computer revolution for various aspects of organizations is taken up and critiqued.

3. THE NATURAL WORLD AND THE ARTIFICIAL WORLD

Basic to Simon's systems-theoretical views is the distinction between two terrains in reality: the natural and the artificial. The natural world is partly separate from the human – think, for example, of a jungle – and partly tied to humans. The artificial realm includes the whole of human culture. The coherence between the two worlds is in the human as well as in what humans bring forth. The human functions using intellectual activities in the artificial world. This artificial or human world is tied, however, to the natural world, through the human brain. Furthermore, human artifacts must obey the same natural laws that natural things obey. In this sense artifacts are not isolated from nature. What particularly distinguishes them from the natural world is that they also derive their identity from the human purposes which they embody.[9]

Generally, we employ the term "artificial" in connection with technology, which we then consider to be a particular sector of culture. Simon takes the world of the artificial in a much broader sense. He strives to impress upon his readers that it is not only the technician who creates artifacts but also, for example, the physician who decides upon a plan of therapy for a patient, or the businessman who develops a sales strategy.[10] Both the medical therapy and the sales strategy, and many more cases besides, belong to the artificial realm designed by human beings. In fact, in Simon's perspective, the artificial world includes all human activities and the whole of human culture. Artificial contrasts with natural and pertains to all that man has made.[11]

Thus what is given naturally is confronted by Simon with what is human-made or has been brought into being through "synthesis." While science is concerned with the analysis of reality, technology aims at synthesis, that is, at designing or composing:

As soon as we introducc "synthesis" as well as "artifice," we enter *the realm of engineering*. For "synthetic" is often used in the broader sense of "designed" or "composed." We speak of engineering as concerned with "synthesis" while science is concerned with "analysis" (*Sciences of the Artificial*, p. 7, italics added).

Following Simon's argument, then, one can describe the artificial world as a "technical world," although Simon himself does not employ this phrase. The artificial or technical world includes the entire world formed by human beings. I have already observed that Simon employs the term "artificial" in a broad sense and that he does not want to restrict it to artifacts such as computers, airplanes, and the like. The concept "technology" undergoes a similar broadening. He does not regard it as essential in speaking of technology that there be *physical* artifacts. The essence of technology is the knowledge, the construction plan or design, underlying the artifact. He accordingly includes social planning, for example, as part of technology.[12]

The word "technology" usually connotes to us the physical artifacts we have created to assist us in our work and in our pleasure: airplanes, television, power plants, automobiles, computers. At a more fundamental level, technology does not consist of things but of the human knowledge that enables us to build and use these artifacts. *Technology resides not in objects, but in knowing (Man and His Tools*, p. 1, italics added).[13]

4. THE SCIENCES OF THE ARTIFICIAL

A key problem that constantly engages Simon is that of how the artificial world, the result of human action, can be charted by science. Again and again one encounters the following difficulty in principle when studying artificial phenomena: they give the impression of being contingent, in contrast to natural phenomena, which reveal a necessary character.[14] This gives the natural sciences a certain edge with respect to the sciences of the artificial. While the former have succeeded in the last centuries in getting to the bottom of natural phenomena, the latter are still practically in their infancy. In this connection Simon recalls the well-known saying of the sixteenth-century Dutch physicist and mathematician Simon Stevin: "Wonderful, but not incomprehensible" (*Sciences of the Artificial*, p. 4).

Simon's meaning is clear: It is the task of science to discover the hidden pattern in the artificial world of man. To carry out this task one will have to begin by defining clearly the domain which is to be the object of the sciences of the artificial. Simon provides four criteria: (1) artificial things arise *via* synthesis by human beings; (2) artificial things may imitate the appearances of natural things while actually differing from them in many respects; (3) artificial things can be characterized in terms of functions, goals and adaptations; and (4) artificial things are discussed in imperative as well as descriptive terms.[15] According to the third criterion, artifacts can be viewed from the standpoint of systems theory as adaptive systems seeking the attainment of certain goals. This characteristic assumes that in artifacts there is a relation between: (1) the *goal*, (2) the inner environment, the nature of the artifact, and (3) the *outer environment*, the surroundings in which the artifact functions. For the theoretical analysis of artificial systems two approaches are possible in principle.[16]

First, it is possible in the case of a given artifact to examine it without considering its goal and adaptation to the surroundings. In such a case the artifact would be discussed in the same way as a natural object, namely, in natural-scientific terms. Simon takes as an example the theoretical discussion of an airplane and a bird. With respect to an airplane, we can describe its inner environment (its power plant, for example), its outer environment (atmospheric conditions), and the relation between the two (the movement of an airfoil) in a natural-scientific way, just as we can with respect to a bird.

> *Given* an airplane, or *given* a bird, we can analyze them by the methods of natural science without any particular attention to purpose or adaptation, without reference to the interface between what I have called the inner and outer environments (*Sciences of the Artificial*, p. 10).

Second, one can investigate the behavior of artifacts not as natural objects obeying natural laws but by the method of systems theory as adaptive systems. Then attention is devoted to "something other" than the inner and outer environments analyzable by natural science. What is it that concerns the sciences of the artificial? Simon uses the term "interface." An artifact can be thought of as an interface between an inner and an outer environment, that is, between the substance and organization of the artifact itself, and the surroundings in which it operates. The interface assures that certain goals are attained through adaptation of the inner to the outer environment.[17]

The analysis of artifacts in terms of adaptive systems results in a so-called *functional* explanation. With this approach it is possible to predict behavior without knowing much about the inner environment of the artifact. For example, if we know that a commercial organization aims to maximize profits, then it is often possible to predict what it will do when economic circumstances change, even though we know nothing about its internal structure. In making such predictions we assume that the internal structure in question places no limits on the adaptation of the artificial system to its altered surroundings. If that assumption is unwarranted, then the system will respond only partly to the so-called task environment or surroundings. Partly it will be determined by the limiting properties inherent to the inner structure of the artificial system.[18]

One has to note that the approach to artifacts thus described and the functional account of their behavior are not exclusive; they apply equally to many biological phenomena. For just as the artificial world embodies human goal-seeking, so the natural world exhibits its own goal-seeking. Simon adduces as an example that many arctic animals have white fur. A functional explanation for this is that in the arctic environment, white is the most suitable color for remaining unnoticed by natural enemies. White animals thus have the best chance of surviving in this environment. By natural selection or some other mechanism an evolutionary process is brought about of adapting to the biological environment.

Extending the analogy with the natural world, one can go on to ask how the adaptation to their surroundings of artificial systems made by humans is realized. Simon is convinced that humans play a central role here, analogous to that of natural selection in evolutionary biology.[19] Reason functions, in other words, as the mechanism of adaptation in the artificial world. As I observed earlier, Simon considers the investigation of human rationality to be the heart of his scientific work, so it is appropriate to turn to this issue.

5. "THINKING MAN" AS AN ARTIFICIAL SYSTEM

Simon's view of the artificial world in a general sense can be regarded as a later philosophical or theoretical expression of the basic idea that guided his scientific investigations of human thought processes from the outset – of the idea, namely, that the thinking human being is an artificial system, of the same type as the computer.[21] Notice carefully that Simon is not guided by the idea that human thought can be viewed *as if* it is artificial; no, human thought *is* artificial. The comparison of human and computer is, in his view, a comparison of two artificial systems of the same kind – namely, information processing systems, which he also calls symbol systems.

The computer is a member of an important family of artifacts called symbol systems, or more explicitly, physical symbol systems. Another important member of the family (some of us think, anthropomorphically, it is the *most* important) is the human mind and brain (*Sciences of the Artificial*, pp. 26–27).

It can be asked what status Simon grants the proposition that human thought is artificial and that computers can think. Are we dealing here with an ontological conception that regards facts in a certain light or are we dealing with a variable scientific statement? Simon emphasizes the latter: the proposition has the character of a *scientific hypothesis*. He formulates it more formally as follows:

Stated a little more formally, the hypothesis is that a physical symbol system of the sort I have just described has the necessary and sufficient means for general intelligent action. The hypothesis is clearly an empirical one, to be judged true or false on the basis of evidence (p. 28).

Now, how can this hypothesis be tested empirically? Simon believes there are two possibilities. One is to develop computer programs and then to investigate their intelligent behavior. The other is to assemble experimental data about the psychology of human thought, in order to ascertain if the human brain functions as an artifact, in this case as a symbol system.

Let us take a closer look at Simon's discussions of cognitive psychology. The central question is: How do people think in solving problems? What tortuous path does their thinking take from a certain starting point (the problem) to a terminus (solution)? Why do people in their thinking not move in a straight line from start to finish? In connection with his hypothesis that human thought has to be viewed as an artificial system, Simon comes to the position that human behavior, in thinking, only gives an appearance of complexity but is in fact very simple. The complexity manifest in the thought process is not inherent in thinking but is a reflection of the complex problem environment. Thus Simon arrives at the following reformulation of his original hypothesis:

A man, viewed as a behaving system, is quite simple. The apparent complexity of his behavior over time is largely a reflection of the environment in which he finds himself (pp. 65, 95).

Simon illustrates this hypothesis with an example. Observe an ant making its way across the uneven surface of a beach. Zigzagging, the ant seeks its way, adapting to the obstacles it encounters en route, sensing infallibly all the while where its nest is. The complexity of the path taken by the ant resides in reality not in the ant but in the surface of the beach. In this case too one can speak of artificial behavior, that is, of the behavior of an adaptive, goal-seeking system.[22] In his psychological studies of human thought, Simon thus proceeds from the assumption that the human being is an adaptive system. In this case the outer environment is formed by the problem – there is a so-called problem environment – while the inner environment consists of the psychological mechanism, the human brain, which enables thought. As an adaptive system a human in his thinking behavior will reflect mainly the characteristics of the outer environment and only a few traits of his inner environment.[23]

To gain empirical verification of the hypothesis in question, Simon and other researchers have carried out psychological experiments with human subjects. They have investigated, for example, the ways in which chess problems and so-called cryptarithmetic problems are solved. It is easy to see that it is beyond the capacity of people confronted with such problems to consider all the possibilities. How, then, is the problem tackled? Experiments show that a human being is capable of employing different strategies, whereby the search through the extensive maze can be reduced drastically – the more sophisticated the strategy, the less search work required to bring the problem to a good solution. The procedure of trial and error is virtually fully replaced by a combination of searching and reasoning. Once the strategy has been established, the course of the search process appears to depend only on the structure of the problem and not to be influenced by one or another characteristic of the problem solver.[24]

Yet it is possible through observation of the problem solver, human or computer, to infer characteristics of the inner environment. From the mistakes made and the capacity to recover from them something can be said about the capacity or preciseness of the memory, the elementary speed of the thought processes, and so forth. From this sort of psychological research into the limitations of the inner environment of humans as thinking behavior systems, it appears that humans have a practically unlimited semi-permanent storage capacity in their memory – for example, one retains certain facts or events throughout one's entire life – but the capacity of working memory is much more restricted.[25] In the end Simon concludes that all the experimental psychological research has corroborated his hypothesis concerning human thought behavior. It is indeed adapted to goal-seeking and is therefore artificial in terms of the previously stated definition. There is therefore no essential difference between human beings and computers as information processing systems.

This conclusion has evoked a great deal of criticism, because it means that the human being no longer occupies a unique place in reality as a rational animal. "Now information processing challenges this last refuge of man's uniqueness" (*Man and His Tools*, p. 15). The philosophical criticism of Hubert Dreyfus (1972) and Joseph Weizenbaum (1976) Simon rebuts with the observation that it is based on a presumptuous view of what it is to be human.[26] Rejecting such a view may not be equated with surrendering human dignity. Human worth does not depend on some unique position human beings might be thought to occupy in reality. Thus, though he begins his reflections with a sharp distinction between nature and artifice, Simon arrives at the idea that the human is fully exhausted in nature:

> We have learned that our human worth does not depend upon our remaining *apart from* nature. On the contrary, the human future depends on our being truly and fully *a part of* nature. Information processing psychology, far from threatening our place in things, gives us further evidence of the closeness of our kinship with all creation (*Man and His Tools*, p. 16).[27]

6. INTERIM ASSESSMENT

Initially we distinguished two main points regarding Simon as a systems thinker: first, the systems-theoretical framework of his thought and, second, his view of the implications of computer technology for society. Before turning to the second point, it will be useful to place a number of critical comments in the margin apropos of what has already been said.

Simon employs the fundamental distinction between two terrains in reality: the natural and the artificial. At first glance the introduction of the term "artificial" appears to be a neutral affair with respect to which no view of reality is yet at stake. But that is only apparently so, as can be seen from three observations.

First, the term "artificial" levels the rich diversity in the human creating and preserving of culture. The examples mentioned, of a plan of medical therapy and of a commercial sales scheme and technical design, may be alike in the sense that all three are the result of human mental activity, but in the cultural sense they are entirely different.

Second, not only does the term "artificial" level all cultural diversity, it brings all human cultural creation into the sphere of technology. The world of the artificial is in Simon's train of thought nothing other than a world that can be constructed in a technical way.

Third, Simon's view puts human thought under the sway of the technical world picture. Thinking human beings are considered in the image of artificial, information-processing systems. Simon believes there is no essential difference between human and computer as information-processing systems because human beings *are* information-processing systems.

Against this view one can make two points: (1) That Simon casts his conception of man and computer in the form of a serious scientific hypothesis strikes the objective reader as somewhat remarkable, to say the least. What facts justify such a far-reaching hypothesis and are clarified by it? There are no such facts, nor does Simon present any. Naturally, as is generally known, there are computer programs which enable people to solve complex problems – for example, chess problems. Yet it is incorrect to conclude from this that the human being, like the computer, is just an information-processing system. All that has been established is that computers are able to imitate a certain part of human thought behavior. But this is not the creative core of human thinking, from which Simon abstracts.

(2) Simon's comparison of humans and computers is based on a systems-theoretical method of analysis. In other words, Simon views the thought behavior of humans and of computers as the behavior of systems which, adapting to their environments, seek certain goals. And the environment is in this case a certain problem environment. What Simon makes no allowance for is that for human beings the problem environment is not an incontestable permanent given. In a certain sense one can say that problems first arise when human beings define them. Human beings possess the unique capacity to impose a particular measuring stick upon reality and to formulate problems in terms of it. In essence, that is the puzzle of human creativity and responsibility, which Simon completely ignores. He addresses human thought only after a problem environment has been defined. That is why his comparison of

the human and the computer breaks down. The problem environment of the computer is already defined and set for it beforehand, in that human beings work with great precision to define it and to develop it in a program. But the thinking human being defines his problem environment for himself. If this point is overlooked, then one is compelled to arrive with Simon at the eradication of the fundamental boundary between human and computer. As might be expected, this view of human and computer works its way into Simon's reflections on the importance of computer technology for society.

7. REFLECTIONS ON COMPUTER TECHNOLOGY

The social consequences of computer technology are compared by many, including Simon, to the far-reaching effects of the steam engine. In this connection there is talk of the first and second industrial revolution, or of the energy revolution and the information revolution.[28] Various sectors of society have already been affected profoundly by computer technology. Experts are of the opinion that we are still at the threshold of an enormous development, the scope of which is virtually beyond comprehension. The dynamics of the field in which computer technology and its applications is developing makes reflection upon it a difficult assignment. Not only do people feel overwhelmed by the developments; they have also to take seriously into account unanticipated new possibilities and unexpected problems. The time and the serenity needed for thorough social preparation for computer technology seem almost entirely nonexistent.

Given this situation, which in the meantime has become characteristic for the whole of modern technology, it must be considered extremely worthwhile if in the first place the experts involved make a contribution to the process of reflection. Their knowledge of the facts and insight into the possibilities can guard us against unfounded optimism or pessimism. Yet according to Simon we need not be surprised that among the experts there is no common opinion. He attributes this to the fact that the social consequences of technological developments are also difficult for the experts to assess. That has been true of earlier technological developments as well. "We only have to look back at the history of the automobile, the airplane, or the radio to see how different is the understanding required to introduce and develop a new piece of technology from the understanding needed to assess its economic and social implications, to anticipate them, and to deal with them" (*New Science of Management*, p. 3).

Since no agreement is to be expected from the experts about the assessment of computer technology, it seems virtually impossible for a layman to form an opinion. Which expert, after all, should we follow? Yet there is a simple yardstick available, Simon says; we choose the experts who on the basis of common sense can make clear how they have arrived at their standpoint. Yet it can be asked whether in practice common sense – what, precisely, is it? – affords such a solid point of departure. Must one not rather infer from the existing differences of insight that the "common knowledge of the world" to which Simon would appeal must not be overestimated, to say the least?[29]

To make his own views clear for the layman, Simon distinguishes three dimensions in discussions of computer technolgy: a technological, a social-economic, and a philosophical dimension. Each of these dimensions has two extremes: the poles of radicalism and conservatism. Furthermore, it also needs to be noted that the philosophical dimension is not entirely different from the technological. While the latter pertains to the possibilities of computers, the philosophical dimension is concerned with human nature. Do human beings possess unique faculties that cannot be equalled by computers?

Simon summarizes his position in the debate about computers as follows: "I am a technological radical, an economic conservative, and a philosophic pragmatist" (*New Science of Management,* p. 6). These three elements of Simon's position mean: (1) computers are able to do the same work that people do (technological radicalism); (2) automation will not drastically change the economic world (economic conservatism); (3) practice will have to show any eventual limitations the computer may have in comparison with human beings (philosophic pragmatism).

8. THE COMPUTER AND DECISION MAKING

In decision-making processes in organizations, the computer is being used more and more at all levels. To make this possible, the decision-making process needs to be subjected to precise analysis. Decision making in Simon's conception involves the following four main phases: (1) determination of the moments of decision; (2) determination of the various possibilities for action; (3) choosing from among the options presented by the second phase; and (4) evaluation of choices made. At the same time, it should be noted that the process is in reality much more complex than a simple sequence of the four phases.[30]

On the broad scale of the problems involved in decision making Simon distinguishes two extremes, which he refers to as "programmed" and "unprogrammed" decision making. Between these extremes is a continuum of transitional types. In the case of a programmed decision, human beings traditionally relied on routine, while at the other extreme they were always guided by creative insight and intuition. The terms, "programmed" and "unprogrammed," Simon has borrowed from computer technology; a program is a detailed prescription indicating a sequence of operations in a computer.

Introduction of the computer has wrought revolutionary changes across the entire spectrum of human decision making since the Second World War. This was true initially in the area of routine, programmed decisions. One must think in this regard of so-called operational analysis or operations research which, although there is no historical connection, can be considered a continuation of the older "scientific management" of Frederick W. Taylor. Computers have made it possible to simulate the behavior of processes on the basis of mathematical models. With the aid of computer simulations it has proven possible to solve complex problems, such as traffic problems, in a scientific way.

Although developments in the area of programmed decision making are still in progress, we have by now been able, Simon believes, to acquire a picture of their scope and significance. This holds in much smaller measure, as he sees it, for developments in the area of unprogrammed decision making, which made its appearance in the sixties. In the meantime, research into the human thought process has shown, Simon ventures to say, that the so-called mystery of thought is based on an illusion. It has turned out to be possible to develop computer programs that solve a problem in exactly the same way people do. In short, the challenging conclusion is: "The secret of problem solving is that there is no secret" (*New Science of Management*, p. 69).

The possibilities that the computer offers for simulating human thought processes are of practical import for organizations in two respects.[31] First, it will lead to the automation of certain unprogrammed decision-making tasks. Where the line will eventually be drawn between automated tasks and those which emphatically require human intervention is difficult to predict. The computer will not always be introduced where technically feasible because it may not be economically advantageous. Second, the computer derives its practical importance not only as a replacement but also as a support for anyone within organizations. In the latter case, human beings working interactively with computers remain fully involved in decision making. The computer in this case provides support functions that aid the process of human thought.

9. COMPUTERIZING THE ORGANIZATION

The introduction of the computer into the organization to aid decision-making processes also has effects on many other aspects of the organization. Most important are its effects on the structure of the organization, on centralization versus decentralization, and on the functioning of authority and responsibility.[32]

9.1. The structure of the organization

An organization of people in a specific cooperative combination can be thought of as constructed on three levels. At the lowest level the basic work processes take place; in a factory, for example, it is at this level that materials are imported and the product is made, stored, and transported. The everyday comings and goings in this lowest level of the organization fall under the management of a middle level, where routine, programmed decision-making processes take place. In the top level, long range policy is worked out on the basis of unprogrammed decision-making processes.

Computerization will in principle not alter this three-level structure of the organization, according to Simon. To be sure, one can anticipate that the relations between the levels will become clearer as a result of an explicit and formal description of the total system. Such a formal description is a prerequisite for introducing the computer into complex decision-making processes. The structure of the entire system must be designed first, whereupon it becomes possible to use the computer at key points.

Besides having a three-level structure, an organization can also be characterized as a whole made up of parts. An organization consists of components, which in their turn are made up of components, and so forth. This whole-parts construction implies a hierarchical structure. Hierarchy is in fact not a specific characteristic of organizations but a characteristic of systems in general.[33] Simon makes the judgment that this hierarchical structure of the organization will not be damaged by computerization. For with the application of computers it will also be necessary to identify the structure of the decision-making process and in so doing to impose a hierarchy on the organization. Nevertheless, computerization can have a significant impact on the other aspects of the organization that have been mentioned. Differences can thereby arise between organizations with respect to the degree of centralization as well as with respect to authority and responsibility.

9.2. Centralization versus decentralization

Centralized decision making is often equated in people's minds with a bureaucratic, authoritarian, and dehumanizing system, while decentralized decision making is viewed as a prescription for personal development of the individual and the humanizing of the organization. Simon correctly rejects such an absolute opposition: "The question is not *whether* we shall decentralize, but *how far* we shall decentralize" (*New Science of Management*, p. 115).

Thus in the first decades following the Second World War there was a clear trend toward decentralization, which went hand in hand with strong corporate growth. In the last decade the trend seems to have turned toward centralization. The latter trend has been further reinforced, according to Simon, by the introduction of the systems approach and computer technology, which have shaped new possibilities for a far-reaching centralization of the entire corporate process.

9.3. Authority and responsibility

In *Administrative Behavior* Simon analyzes organizations in terms of decision-making processes. He views an organization as a system that interconnects the decision making of its members. Within an organization the decision-making behavior of one person is linked to that of others. Those who have the power to control the behavior of others in the organization use it to exercise authority, according to Simon. And he defines authority in the following terms:

> "Authority" may be defined as the power to make decisions which guide the actions of another. It is a relationship between two individuals, one "superior," the other "subordinate." The superior frames and transmits decisions with the expectation that they will be accepted by the subordinate. The subordinate expects such decisions, and his conduct is determined by them (*Administrative Behavior*, p. 125).

Or more briefly: "Authority is exercised whenever a person allows his decisions to be guided by decision premises provided to him by some other person" (*New Science of Management*, p. 96).

The criticism that authority is dehumanizing by nature, because it may not be compatible, for example, with the self-realization of the individual, is controverted, according to Simon, by practice, which shows just the opposite. People have been shown to feel ill at ease and to be hindered in their development if there is no structure in their surroundings. The organization offers the best circumstances for self-realization when it achieves the right balance between freedom and the limitation of freedom. Where precisely this balance must lie is of course not easy to say, but Simon is certain that the experience of freedom and responsibility does not require total independence from one's surroundings. To make this clear he draws a comparison with the situation of a sailor who must constantly react to the changing circumstances of the wind and the waves. "One does not feel unfree handling a sailboat, even though most of one's responses are governed by the moment-to-moment demands of wind and wave" (*New Science of Management*, p. 121).

This example of the sailor illustrates yet another point: It makes no difference according to Simon whether man's freedom is restricted by human power or by other factors in his surroundings. With respect to the computerization of the organization he is therefore of the judgment that human authority relations can be transferred without objection to a rational-technical-organizational system. As automation progresses, such a system will increasingly function without human intervention. The task of the manager of the future will ultimately consist of guarding and improving automated systems. "Line management, as we have known it in the past, will be a smaller part of the job, and managers will spend much of their time and effort as members of task groups engaged in analyzing and designing policies and the systems for implementing them" (*New Science of Management*, p. 123).

FINAL ASSESSMENT

My interim assessment argued that Simon's thought is governed by the technical world picture. This is also clear from the standpoint he adopts in the debate about computer technology. Simon is, as he says, a technological radical and a philosophic pragmatist. Here it must be asked critically whether practice is able, as he believes, to serve as the referee between the extreme positions of philosophic conservatism and radicalism. In other words, can the on-going actual development of computer technology indeed decide the question of whether there is or is not a strict boundary between human possibilities and those of the computer?

Simon proceeds here on the incorrect assumption that practice is an independent affair. Practice, in this case the development of computer technology, is always judged from a particular perspective and developed further in the light of that perspective. If the radical believes that human beings are not unique and that computers are able in principle to do all human work, then practice will not be able to disprove him. When the computer cannot do certain things, the defense can always be that future developments will solve the problem. It should therefore not be surprising that Simon has never withdrawn his presumptuous and confident assertions of the sixties. Practice will never be able to convince him of error; at most it will be able to disillusion him. But Simon's faulty view of the human does have consequences for practice; it adapts to the technical world picture. The anthropomorphic view of the computer and the computeromorphic view of humanity cannot fail to be reflected in the process of technicization.

Consider again Simon's view of freedom and responsibility in the organization. Simon advocates a functionalist view of authority. Authority is a necessary function of the organization, such that some have the power to make decisions while others must obey. According to this conception a certain inequality of power is essential for a relation of authority. Authority implies a hierarchical relation between two persons, a relation of chief and subordinate.

Proceeding from a mechanistic view of human behavior, Simon introduces a reduced conception of authority at this point.[34] Within the technical world picture, which governs all his thought, authority can only be a technical-organizational or mechanistic regulating function within the organization. It is on the basis of this view of authority that Simon opposes the notion that authority stands in the way of the self-development of the individual. Self-development as he sees it depends upon a certain limitation of individual freedom, whereupon the idea of freedom also acquires a functionalist content. Freedom in that case, as one will find elaborated more extensively in the writings of the systems philosopher Ervin Laszlo, is reduced to the cybernetically regulated tolerance of the individual within the limits of a system.[35] And these limits are determined by various factors, including that of authority as the power of technical control.

Although Simon correctly understands the indispensable role of authority for human society, his view of it fails. Because he proceeds from a reduced view of authority, freedom, and responsibility, Simon has no objection, upon introducing the computer, to replacing personal relations of authority in the organization with an impersonal technical-organizational structure. Therefore the deeper cause and meaning of the criticism that is brought to bear against the dehumanizing character of authority construed as power also eludes him.

Free University, Amsterdam

NOTES

1 Translated from Dutch by Herbert Donald Morton.

2 Sytse Strijbos, *Het Technische Wereldbeeld: Een Wijsgerig Onderzoek van het Systeemdenken* [The technical world picture: a philosophical investigation of systems thinking] (Amsterdam: Buiten and Schipperheijn, 1988). Includes an English summary.

3 "Culture" is here taken in the current broad sense; it can be defined as "whatever gives form to existence." It would perhaps be more precise to distinguish between the activity that forms a culture and its results. In that case, the broad meaning would require a better definition: "Whatever forms or gives form to existence, including the result or outcome."

4 The renowned British cultural historian, Arnold Toynbee, presents a fascinating account of the meeting of Western with other cultures, including the accompanying process of technological global integration. See his *Change and Habit: The Challenge of Our Time* (London: Oxford University Press, 1966).

5 E.J. Dijksterhuis, *The Mechanization of the World Picture* (Oxford: Oxford University Press, 1961; Dutch original, 1950). In the mechanistic world picture, the artificial, technologically formed object serves as the model for reality as a whole. *Mechane* means artificial. One can, therefore, equally well speak of a *technical* (or technological) world picture. That is in fact what I prefer, since not every technical object is mechanical.

6 I elaborate on this in my book cited in note 2, above. For Ackoff, see his *Redesigning the Future: A Systems Approach to Societal Problems* (New York: Wiley, 1974); and for Boulding, see his *The Image: Knowledge in Life and Society* (11th ed.; Ann Arbor: University of Michigan Press, 1977; original, 1956).

7 Herbert A. Simon, *Administrative Behavior: A Study of Decision-Making Processes in Administrative Organization* (2d ed.; New York: Free Press, 1976; original, 1947).

8 Throughout the remainder of the paper, I will use short titles of Simon's books in the text, with page citations in parentheses.

9 Simon, *The Sciences of the Artificial* (2d ed.; Cambridge, Mass.: MIT Press, 1982; original, 1969), pp. 5 and 6.

10 *Ibid.*, p. 129.

11 *Ibid.*, p. 6.

12 *Ibid.*, chapter 5, "Social Planning: Designing the Evolving Artifact."

13 See also Simon, *The New Science of Management Decision* (3d ed.; Englewood Cliffs, N.J.: Prentice-Hall, 1977; earlier editions, 1960 and 1965), pp. 165ff.

14 *Sciences of the Artificial*, pp. ix–x.

15 *Ibid.*, p. 8.

16 These two approaches to the theoretical analysis of artificial systems agree on the distinction within systems theory between structural and functional descriptions of systems. Von Bertalanffy formulates the distinction this way: "Internal description is essentially 'structural,' i.e., trying to describe the system's behavior in terms of state variables and their interdependence. External description is 'functional,' describing the system's behavior by its intervention with the environment." See Ludwig von

Bertalanffy, *Perspectives on General Systems Theory: Scientific-Philosophical Studies* (New York: Braziller, 1975), p. 163.
[17] Simon, *Sciences of the Artificial*, pp. 9 and 132.
[18] *Ibid.*, p. 16.
[19] *Ibid.*, p. 11.
[20] Simon takes up the analogy between rational adaptation and evolution as a separate subject in *Reason in Human Affairs* (Oxford: Blackwell, 1983), chapter 2, "Rationality and Teleology."
[21] The view of human beings as artificial systems, according to Simon, is applicable not only to cognitive but to motivational, including emotional phenomena. See *Sciences of the Artificial*, p. 65, note, and *New Science of Management* (note 13, above), pp. 16 and 35.
[22] *Sciences of the Artificial*, p. 63.
[23] *Ibid.*, pp. 62 and 97.
[24] *Ibid.*, p. 71.
[25] *Ibid.*, pp. 72–77. In the context of his theory of human thought, Simon also devotes attention to linguistic abilities. Language, according to him, must also be regarded as an artificial construction. On this, see *Sciences of the Artificial*, pp. 89–95.
[26] *Ibid.*, p. 97; see also *New Science of Management* (note 13, above), pp. 3–4. For Dreyfus, see his *What Computers Can't Do* (New York: Harper & Row, 1972); for Weizenbaum, see his *Computer Power and Human Reason: From Judgment to Calculation* (San Francisco: Freeman, 1976).
[27] See also *New Science of Management*, p. 37.
[28] *Ibid.*, p. 166; see also Simon, *Man and His Tools: Technology and the Human Condition* (Amsterdam: Duijker-Lezing, Intermediair Bibliotheek, 1981), pp. 9–10.
[29] *New Science of Management* (note 13, above), p. 8.
[30] *Ibid.*, pp. 40ff.
[31] *Ibid.*, p. 47.
[32] *Ibid.*, pp. 110–123.
[33] See Simon, "The Architecture of Complexity," *Proceedings of the American Philosophical Society* 106 (December): 467–482.This article, frequently cited by people writing on general systems theory, is reprinted as chapter 7 in *Sciences of the Artificial.*
[34] See Alasdair MacIntyre, *After Virtue: A Study in Moral Theory* (London: Duckworth, 1981), pp. 25–26 and 79–80, where a similar criticism is raised.
[35] See Ervin Laszlo, *The Systems View of the World: The Natural Philosophy of the New Developments in the Sciences* (New York: Braziller, 1972).

LADISLAV TONDL

CHANGES IN COGNITIVE AND VALUE ORIENTATIONS IN SYSTEM DESIGN

The phrase, "cognitive and value orientations," when applied to technology, presupposes that a technological system is a materialization or instantiation of a set of cognitive features and a set of valuations. Discussions dealing with the explanation of technological change commonly point out the role of value orientations and the significance of decision making as influenced by accepted values. Discussions of new cognitive orientations in technology generally call for an interdisciplinary approach involving the dynamics of behavior, spatial and temporal impacts, and prognoses with respect to the preparation for, the planning of, and the design of particular technological processes. New approaches in value orientations postulate that technical and humanitarian rationales can be combined; people speak of "human dimensions" or of the "controllability" or "manageability" or "acceptability" of technologies, or else they talk about solving "ergonomic" or "ecological" problems associated with technology. This essay is an attempt to clarify these changes.

1. THE TERMS, "COGNITIVE ORIENTATION" AND "VALUE ORIENTATION," AS APPLIED TO TECHNOLOGICAL SYSTEMS

Some recent Soviet discussions of the outlook for technical disciplines (Gorokhov, 1982), including system design and social/cultural engineering (Rozin, 1984; Sidorenko, 1984) – discussions that also have as their background the experience of recent technological accidents, of the increasing likelihood of hazards and other negative impacts of particular technological developments – have focused attention, among other things, on two important interconnections. These tend to be called the "cognitive" and "value" orientations of technology or techno-engineering thought processes. The phrases are intended as generalizations and modifications of concepts stimulated by the pioneering work of Thomas S. Kuhn on the role of paradigms in scientific and technological development. In the usage of these Soviet writers, any scientific/technological artifact, anything new that results

Paul T. Durbin (ed.), Broad and Narrow Interpretations of Philosophy of Technology, 87–98.

from rational cognition or creative activity, should be characterized in terms of the knowledge claims or values projected into its creation. In these terms, any novelty in science or technology should be characterized as "theory-laden" (or "knowledge-laden") and "value-laden."

It seems obvious that the increased complexity of technological systems – as they are envisaged, planned, and designed – involves an increased range of knowledge at each stage, not only design, but also realization and implementation. The synthesis, aggregation, integration of the required knowledge – bits of which, as a rule, come from different sources, from a variety of scientific and technicalines, and which include knowledge not only of what can be realized but how, under what conditions or limits – all of this requires a systems approach or a systematic analysis. As a result it is no wonder that the design of large-scale technological systems involves the management of organized teams made up of specialists from different fields – and not just scientific and engineering disciplines but also economists, biologists, medical specialists, and (most important) ecologists ("Inzhenernaya deyatelnost i nauka," 1986; Entwurf für eine Richtlinie, 1988).

While the cognitive orientation and associated prerequisites of the conceptual preparation for the design of a technological product may seem obvious – something corroborated by the fact that this conceptual preparation for designing is always carried out by a specialist or teams of specialists – the situation is entirely different where the value orientation is concerned. Nonetheless, the origin of any technological product or process must inevitably involve particular evaluations and decisions, beginning with the assessment of extant needs, obligations, or requirements as acceptable or justified, and ending with the choice of a particular technical solution for the particular problem. These evaluation and decision processes always involve particular values, priorities, and criteria. This necessary characteristic of the origins of technological products remains necessary even when such values and decisions do not need to be formulated or expressed explicitly – even when they are concealed. Such evaluation and decision processes also reflect particular interests or demands of those involved in carrying out technological projects – another reason for saying evaluations and decisions are a necessary characteristic.

The value orientation of technological projects – envisaged, planned, or designed – may, as several Soviet authors have recently indicated, at times be subject to a certain sort of contradiction. On one hand, it is rightly expected that a team of specialists will submit a solution that is optimal in terms of meeting the demands of the society as a whole while, at the same time, it complies with the latest developments in science and technology, including the most up-to-date knowledge. On the other hand, any proposed technical solution must comply with the needs and interests of the customer – and perceived needs and interests may be narrowly biased, based on limited knowledge or limited value horizons; or they may be the result of outdated ideas or concepts, or of a single-interest and partial approach to technical problems that amounts to a bias. It may be assumed that this sort of contradiction is at least partly responsible for some of the deficiences still to be found in the area of investments; for the fact that technological development still tends to favor existent and sometimes even outdated technical paradigms, i.e., the repetition and continuance of previously accepted models, standards, designs, and technological methods; for increasing delays of technological projects – in other words, for inadequate attention being given to the role and importance of new technical directions.

2. TECHNOLOGICAL TRENDS AND A PROPOSAL FOR A "TECHNOLOGICAL DIAGNOSTICS"

In the past, analyses of technological trends and technical developments were carried out mainly using the term "innovation" introduced by Joseph Schumpeter. Authors focused their attention principally on the role and influence of technical innovations in economic development, on typologies of technical innovations, on so-called innovation cycles or rhythms, etc. Along with these went other analyses – of so-called "technological arrest," "technology transfer," the importance of "innovation clusters" (i.e., centers of scientific and technological creativity) – that were carried out in Western Europe and Organization for Economic Cooperation and Development (OECD) countries. Such analyses have provided a great many worthwhile incentives especially in terms of directing science and technology policies or programming and planning technological developments that show promise. (Although – despite a few initial promising efforts – the value of analyses of this sort has been badly underestimated in our country, leading to their discontinuation here, they remain worthwhile.)

More recently, analyses of problem spheres different from those just mentioned, but that represent a certain expansion of them, have been initiated. Where the earlier efforts focused mainly on scientific and technical innovations, on their genesis, or on the technical, informational, or social prerequisites of developing them fruitfully, the more recent efforts have more commonly discussed problems related to a new and more general phrase, "technological change." The essential differentiation of the latter from analyses that use concepts associated with the term "innovation" can be characterized simply. "Technical innovation" tends to emphasize changes that are required or desirable in the light of current technical possibilities and are generally assumed in advance to be beneficial – or, to be more precise, are assumed to lead to a reduction in entropy or disorder in processes in which humans control, manage, or utilize transformations in materials, energy, or information. By contrast, "technological change" equally well covers undesirable changes, changes that involve negative impacts (e.g., on the environment) or increased risks or an increased likelihood of accidents. Clearly this is not to say that the ratio of positive to negative impacts of technology is constant, that it cannot be altered to pursue more desirable directions or new technological changes of a more appropriate sort.

Issues associated with the phrase "technological change" can be examined from a variety of viewpoints with different consequences. Such topics include the main trends and directions of technological development, their assumed or possible trajectories (typically expressed in tree or branching models), and the plotting of these directions according to temporal or spatial dimensions (this being little more than an expansion or generalization of earlier "technology transfer" analyses). Here one important problem area needs to be mentioned – an area that has recently received much attention under the heading, "explaining (or diagnosing) technical change."

To explain technological change in this new broader sense requires analyses of the origins and of the course of particular technological developments – typically, the reconstruction of causal links leading to a technological change of a particular sort. In such analyses, attention is paid chiefly to general relationships or to dependencies that have a repeatable character – so that, assuming identical or at least analogous conditions, an identical or analogous technological change can be predicted (Tondl, 1986b). For this reason, the approach is sometimes called "technological diagnostics." (In formal terms, explanations and diagnostic models have similar or analogous structures: each is represented in a usage that employs general data – scientific laws, hypotheses, or other sorts of generalizations – particular data facts, and inference rules. Hence the similarity or analogous meaning between "technological diagnostics" and "explaining technological change.")

"Explaining technological change" – using the phrase as a synonym for "technological diagnostics" – is important in analyzing especially those past changes which can be characterized as undesirable, as unwanted or unexpected, as having negative impacts or making matters worse or increasing risks or any other dangers however characterized. The fact that we can explain such changes, tracing out the sequence of processes that led to the particular sort of technological change (following the tree structure down the branches to the trunk to the roots), is especially important if we are to take measures to prevent, or at least reduce the probability of, such undesirable changes or negative impacts in similar situations in the future. For all practical purposes, the function of explanation in such cases is the same as diagnosis – and a diagnosis is useful or satisfactory if it is capable of initiating effective therapy (Tondl, 1974).

However, explaining technological change (in this sense) is by no means limited to past changes and negative impacts. It can also focus on envisaged changes, changes in preparatory, planning, or design stages. Since every technological project is the result of human technical activities – made up (usually) of complex, sequenced, hierarchically organized and interrelated sets of subactivities, especially cognitive activities but also proposals, evaluations, decisions, and implementing activities – a satisfactory explanation of any particular technological change which includes a knowledge of all the factors must be detailed indeed. Unlike explanations of phenomena or situations or occurrences that do not require, for their appearance, any human intervention – that is, natural relationships and causal connections in the strict sense – explanations of technological changes must take into account that they are the result of human purposes and organized human activities which involve a broad range of subjective phenomena. Among other things, the range includes expressions of needs, demands, wishes, etc. This includes, of course, in addition to the "value orientations" mentioned earlier, "cognitive orientations" as well.

These subjective elements in the diagnostic explanation of technological change can, in conclusion, be summarized as follows:

(i) Every particular case of technological change can be analyzed in terms of a conscious, organized, goal-directed set of activities which can be transformed, making better uses of materials, energy, or information (Pejsa, 1986; Tondl, 1986b).

(ii) Every technological change is artificial ("second nature"); that is, represents conditions, occurrences, processes brought about by practical human activities (Tondl, 1987b).

(iii) Every technological change that is a result of planning or human requirements and designs had to have been – before it was put into execution – first proposed as an "ideal model" (or an "information model") or project, including what was to be changed, how it was to be changed, and for what purpose (Tondl and Pejsa, 1987). (This third summary item stresses the indispensability of a preparation stage in any planned or targeted technological project.)

3. CHANGES IN COGNITIVE ORIENTATION IN DESIGN

Changes in the cognitive or conceptual orientation of technological design – mainly in the preparatory stages of research, design, and development – are today motivated chiefly by the need for a systems approach at every stage: preparation, implementation, and use. That is why technological design is often called "system design" today (Gasparski, 1978; Gorokhov, 1982).

(It should be noted parenthetically that this term is not always used unambiguously. It can denote either the use of system principles and the methods of systems-engineering in technological design projects – see Tondl, 1987a – or else the design of systems as such. Nor is there any clear dividing line between the two. In what follows, I will be using "system design" in the first sense.)

The main modifications today in cognitive orientation for engineering design – modifications reinforced at every step and which *must* be taken into account in every sphere of engineering and technical activity – can be summarized as follows. There must be (a) provision for interdisciplinary cooperation and an interdisciplinary synthesis; (b) emphasis on a systems approach in the preparation, implementation, and utilization stages of technological projects; and (c) application of criteria of prognostication.

Interdisciplinary syntheses have become fairly well established in some technical fields. Moreover, almost all large engineering projects – electrical power stations, large-scale construction for industry, etc. – must necessarily involve the carefully organized and controlled cooperation of a variety of technical disciplines: various engineering specialties, materials science, control-process disciplines, electronics and microelectronics, etc. And in the most recent past these disciplines have been complemented by information science and theoretical and applied cybernetics – with the latter including automatic processes using algorithms, scientific and other databases, and even some features of artificial intelligence.

Many technological projects or solutions of technological problems today require that interdisciplinary syntheses include information about ergonomics – i.e., about human-technological interrelationships, ecological impacts, medical consequences, etc. Solutions of a great many technological problems will be less than complete or viable if they do not take into account knowledge from non-technical fields, not just industrial (or human-factors) engineering but others as well. Some examples of interest here might include recent developments in linguistics, humanistic approaches to the theory of language, and graphic communications – with progress in these fields coming in close parallel with the latest programming developments using the latest generation of computers.

The second requirement, emphasis on a systems approach (even, one might say, a "systems attitude"), is underscored by the fact that today any large technological project – large manufacturing systems, energy plants, transportation systems, planned urban developments – must be conceptualized as a subsystem within a larger system (technoeconomic, socioeconomic, etc.). This requires recognition of complex interrelationships or interactions among individuals in various subsystems, at different times and places; and these must be factored in as early as the conceptual or preparatory stage, before designing starts. The systems approach requires analyses of the dynamic behavior of subsystems, including expected or potential changes in the interactions among subsystems. Such analyses typically include behavior even in extreme situations (e.g., of conflict, situations characterized by increasing risk levels, etc.); and they must pay very careful attention to a whole broad range of anticipated or potential impacts (including the extreme situations just mentioned). Because of the broad range of these potential impacts, a systems approach must focus on ecology or the environment (see Boese, Gutsch *et al.*, 1976) – e.g., on possible changes in climate that could represent a risk to the living environment or biosphere. For all these reasons, most advanced interdisciplinary design teams today include specialists in environmental engineering, and an indispensable part of any large-scale project is models or simulations of expected or potential environmental impacts.

Prognostications – and the application of prognostic criteria – are also an indispensable part of systems designing for large-scale technological projects. The idea, once common, that documentation for a project design need contain a model for the project only at the time it is put into service must now be regarded as outmoded and unsatisfactory. System designs must include anticipated impacts of projects, including their expected dynamics, behavior over time and in different locations, cycles or rhythms, whether they have done what they were supposed to, and so on.

Time parameters represent a special problem and time dimensions in a systems analysis should not be limited to the traditional assignment of a series of numbers at various points in time. A systems analysis must include features associated today with the concept of "temporal directions." This includes not only direction in the usual sense but also the rate of change in speed of the increasing entropy of a system. Other temporal parameters include length of useful life, measures of diminishing reliability, even predictable changes in moral climate or public attitudes.

Since complex technological systems typically involve numerous subsystems, one feature of which are differentiated measures of these "temporal directions," different time parameters related to cycles, etc., an important part of system design is what I would feel obliged to call "flow chart coordination." This is very important in assessing the connectedness or relationships of a technological system to its surroundings, as well as in assessing the relationships of individual subsystems to the broader system of which the technological system being designed is to become a part.

System design, if it includes time parameters or "temporal direction" measures or variables within a model of a technological project being planned, cannot avoid a number of problems – especially if we are talking about investing large sums in the project. These problems include the possibility of needing to build in plans for future updating or modernization or other future modifications once the system is working; problems of secondary or tertiary uses; and, last but not least, problems associated with phasing out or decommissioning or liquidating the project or plant at some time in the future. We should never become careless victims of a naive optimism, overconfident that future generations will take care of these problems – that they need not concern us at all. We know already that such problems exist, for instance with respect to nuclear power plants or (in a different way) large housing complexes made of precast concrete.

Problems with these time dimensions cannot always be determined precisely or quantified exactly, using present knowledge. Technological forecasting nowadays must make use of methods of prognostication based on the views or opinions or recommendations of teams of experts, often enough assembled on an *ad hoc* basis.

4. NEW VALUE ORIENTATIONS

Value orientations in the preparation stages – concepts, designs, research, developments – of large-scale technological projects can also draw in a positive way upon past traditions, including methods and procedures that have been around a long time in technological decisionmaking. For example, we should not forget the methods and criteria developed for the analysis of economic efficiency with respect to investment strategies, or cost-benefit concepts, or so-called "utility functions," multifactorial decision processes, etc. Beginning with the classical work of Wilfredo Pareto at the beginning of this century, a rather rich pool of methods has been available; furthermore, these methods have been effectively employed in technical and technoeconomic decisionmaking.

We can conclude that these methods have lost none of their importance as new solutions are required for new groups of problems under the changing value orientations of contemporary system design. Rather, the methods have been expanded, generalized to a certain degree. The sphere of problems has expanded at least in part because of a new mode of technological thinking called "technology assessment" – for which activities I would prefer the complex phrase, "systems assessments of technological innovations requiring large investments of resources." For now, I would like to show, very briefly, how a systems assessment expands existing means or methods in the area of value orientation; to show, that is, the features the approach accentuates and the trends it promotes.

The changes or new accents in value orientations in systems design that are called for can be summarized in terms of *generalizing* traditional concepts of efficiency and utility functions; *synthesizing* technical with non-technical rationales; and *accentuating* "controllability," "manageability," and the "human dimensions" of new technological developments.

To generalize the traditional concept of efficiency requires mainly that we assess technological projects as purposively conceived subsystems of broader systems, emphasizing the dynamics of system behavior, as well as impacts in extreme situations in both temporal and spatial dimensions. Since methods using the concept of efficiency presuppose definite inputs and outputs for any given technological system, including the functions the system is capable of, generalizing the traditional concept of efficiency requires synthesizing analogous features of the broader system of which it is a part. Similarly, the concept of "utility functions" must extend beyond the immediate users of a particular system and include universal criteria for the long term.

Synthesizing technical and non-technical rationales means principally expanding the sets of technoeconomic criteria used so far to include criteria associated with the protection and preservation of a healthy environment for living things; with biological, medical, psychological, and social perspectives; with cultural and aesthetic values; and with universal moral responsibilities – which includes more than responsibilities that are controllable or manageable. Recent technological development proves beyond a doubt that such issues, for years left only to science fiction writers, are by no means fictional.

5. BASES FOR DECISIONMAKING AND THE THEORETICAL SUPPORT FOR TECHNOLOGICAL SOLUTIONS

The schema for explaining or diagnosing technological changes (demanded, planned, or designed) that I have just proposed presupposes that all technological changes are based on human decisions. Such decisions always involve cognitive and value orientations. With respect to the role of cognitive and value orientations in choosing what turns out to be the ultimate technological solution to a given problem, certain analogous terms used in determining and assessing the prerequisites of a system in terms of *information* – explanation, prediction, diagnosis, and other terms utilizing theoretical and factual data; i.e., databases and theoretical bases for decisions – can be extended to determine and assess the informational values of that particular solution. (I have introduced the term "conditional credibility" to cover the same features of the assessment of scientific projects that "theoretical support" covers here.)

"The bases for decisionmaking" means that set of informational prerequisites necessary to solve a given problem the solving of which requires certain kinds of knowledge (Tondl, 1986a). In science, the decision base is a well-defined pair made up of a set of generalities – scientific laws, hypotheses, and empirical generalizations – coupled with a set of facts. The resultant solution is a derivative (in the technical sense) of the bases for the decision. In a design process, the bases for decisions are typically far more complex; they include:

– the specific characteristics of the given problematic situation, including specifications of needs, requirements, demands, and objectives of the proposed technological solution;

– sets of empirical data, including an adequately detailed description of the initial state-of-the-art, a factual description of the problematic situation, and the surroundings of the system, both geophysical and anthropological;

– sets of general scientific and technical knowledge needed for any possible solution of the given problem;

– the set of all the other conditions necessary for the solution of the problem, including not only such feasibility conditions as adequate resources and technical feasibility but also the standard requirements for any technical solution; and finally

– the set that characterizes the value orientation: all the values, priorities, and criteria needed to arrive at an optimum solution.

The term "support" (where we should recall my analogous term, "conditional credibility," for scientific decisions) refers not only to how but also to the extent to which the resultant solution corresponds to all of its available informational prerequisites. For this reason, the concept of support has become very important in evaluating the results in all systems thinking, where appropriate bases, plus appropriate inferences, assure the right decisions – that is, in all cognitive systems, expert systems, administrative systems, or any other systems using features of artificial intelligence. It is correspondingly important to know, when assessing a design process, how the designer or design team made use of all the information background available – information falling under both the cognitive orientation and value orientation categories.

To conclude, assessing the theoretical support for different alternatives for the solution of a particular technological problem is an important task for those concerned with the systems assessment of design activities in technology. It should be obvious that the interdisciplinary team needed to carry out this systems assessment needs to have all the information necessary to arrive at an optimum solution. But this is just another way of saying that both "cognitive orientation" and "value orientation" are necessary for any good solution of a technological problem.

REFERENCES

Boese, P., R. Gutsch *et al.*, *Planungsfaktor Umweltschutz* (Grafenau: Lexika Verlag, 1976). In German.

Entwurf für eine Richtlinie, "Empfehlungen zur Technikbewertung," in W. Bungard and H. Lenk, eds., *Technikbewertung* (Frankfurt: Suhrkamp, 1988). In German.

"Inzhenernaya deyatelnost i nauka; Materialy kruglogo stola" [Engineering Activity and Science: Round Table Materials], *Voprosy Filosofii*, (1986) No. 5, pp. 71–87. In Russian.

Gasparski, W. *Projektowanie: Koncepcyjne Przygotowanie Dzialan* (Warsaw: PWN, 1978). In Polish.

Gorokhov, V. *Metodologicheskij Analiz Sistemotekhniki* (Moscow: Izd. Radio i Svjaz, 1982). In Russian.

Huning, A. "Zentrale Werte im technischen Handeln," in H. Marhenkel and W. König, eds., *Technikbewertung, Wärmepumpe* (Düsseldorf, 1985), pp. 7–15. In German.

Pejsa, J. "Explanation of Technological Changes and the Question of Efficiency of Technological Changes," *Design Methods and Theories* **20**:4 (1986):519–531.

Rozin, V. "Projektowanie kak Object Metodologicheskogo Issledovania," *Voprosy Filosofii* (1984), No. 10, pp. 100–111. In Russian.
Sidorenko, V. "Genezis Projektnoj Kultury," *Voprosy Filosofii* (1984), No. 10, pp. 87–99. In Russian.
Tondl, L. *Scientific Procedures* (Dordrecht: Reidel, 1974).
Tondl, L. "Some Methods of Information Evaluation of Scientific Results," *Computers and Artificial Intelligence* **5** (1986a):385–394.
Tondl, L. "Classical Explanation and the Explanation of Technological Changes," *Design Methods and Theories* **20**:4 (1986b):506–518.
Tondl, L. "Nova Paradigmata Projektowani a Funkce CAD," in *Proceedings of ICCADI 87 Conference* (Karlovy Vary, 1987a). In Czech.
Tondl, L. "Explaining Technological Changes and the Paradigms in Technology Design," in *Proceedings of the VIIIth International Congress of Logic, Methodology, and Philosophy of Science* (Moscow, 1987b).
Tondl, L., and J. Pejsa. "Metodologicheskie Aspekty Sistemnogo Projektirovanija," *Voprosy Filosofii* (1987), No. 2, pp. 87–96. In Russian.

PART IV

HISTORICAL, CULTURAL, AND POLITICAL CRITIQUES

ANDREW FEENBERG

DEMOCRATIC SOCIALISM AND TECHNOLOGICAL CHANGE*

THE CONCEPT OF SOCIALISM

As the Communist world enters into an ever deepening crisis, socialist theory increasingly detaches itself from practical politics and takes refuge in normative considerations.[1] This development has the merit of clarifying the fundamentally democratic inspiration of socialist ideals, but it runs the corresponding risk of recasting these ideals in a utopian mold that invites dismissal on the grounds of impracticality. Contemporary arguments for democratic socialism are especially vulnerable to the charge that they fail to come to terms with the problems of technology, administration, and the related complex of cultural and educational issues that are ritually brought forward as fundamental obstacles to economic democratization.

In this article, I propose to address these problems of technology and culture in terms of a very broadly reconceptualized theory of the "transition to socialism." My goal is to raise the problem of feasibility to the level of a question of principle, rather than treating it as a laundry list of familiar objections of detail. I believe a discussion of this kind can significantly advance our understanding of the normative questions as well by relating them to the framework of social transformations they are intended to guide.

There is a precedent for this emphasis on the practical framework for the realization of socialism. Marx believed his most important discovery was the idea of a transition linking the capitalist present to a very different future through a specific dynamic of change. Instead of treating socialism as an "ideal," he asserted the existence of a historical "process" or "law" leading from capitalism to socialism.[2] Today it is clear that such a "law" will apply only contingently to any real society, but in any case its logic can be formulated as a general model of the dynamics of cultural change in societies making the transition to socialism.[3] The transition to socialism might then be reconceptualized non-deterministically as a *civilizational project* rather than as the result of what Marx once called a social "process of natural history."[4]

Paul T. Durbin (ed.), Broad and Narrow Interpretations of Philosophy of Technology, 101–123.

Civilizational projects are realized through possible *trajectories of development* that impose a global pattern of culture based on new values that eventually define an entire epoch of human history. Capitalism supports just such a civilizational project, underlying the differences between national regimes. The Marxian model of socialist transition can be employed to define a socialist civilizational project, fundamentally different from that of captalism. Unlike an idealized socialist utopia, which necessarily plays the role of unattainable "ideal" in opposition to the sorry state of social "reality," a dynamic transitional model of the sort proposed here can be used to develop concrete proposals for change.

If socialism is a civilizational change, that would explain why Marx did not treat individual components or aspects of the future society he anticipated as so many policies that need only be implemented by a sufficiently determined government. Instead, he imagined a coherent transformation in the very foundations of the existing social world leading to the creation of a new type of industrial society. The centerpiece of this new society was to be a radical advance in the cultural level of the labor force. It is not easy to reconstruct Marx's view of this transformation, but I will argue that he regarded *socialization, democratizaton,* and *innovation* as intrinsically connected dimensions of it.[5] Here is a brief sketch of these three dimensions:

1. the socialization of the means of production, accompanied by the early substitution of planning for markets in the allocation of capital and other productive forces, and eventually, at a later stage, the general replacement of the market;

2. the radical democratization of society through an end to the vast economic, social, and political inequalities characteristic of class societies;

3. the innovation of a new pattern of technological progress leading to an end to the sharp division of mental and manual labor characteristic of capitalism and a requalification of the labor force.

Any concept of socialism based on these premises can legitimately be called "Marxian" in inspiration. By the same token, the reconceptualization of socialism on the basis of the first or second component alone leads to a variety of non-Marxian socialism. The existing Communist societies would have to be counted among these latter given their narrow overemphasis on planning at the expense of democracy and technological change. Elite control of industry, such as Stalinism presupposes, requires precisely the division of mental and manual labor and the corresponding technological forms the Marxian transition to socialism is supposed to overcome. Similarly, a socialist position such as that of Jürgen Habermas, which reduces socialization to democratization while conceding the neutrality of technology, falls outside the Marxian framework.[6]

This unified conception of socialism has by now been split into its component parts by history and analysis. Popular movements for democracy in socialist countries, like environmental and labor movements for technological change in the West, testify to the breakdown of the original Marxian synthesis, and the need to articulate relatively independent struggles in every sphere. Contemporary social theories which, starting out from Marxist premises, move toward recognition of this fragmentation of socialism are sometimes called "post-Marxist." Such theories attempt to recover the democratic or technological dimensions of socialism against the exclusively economistic Soviet model.[7]

This article presupposes many of the results of this general critique, and applies them in the technological domain. The argument hinges on the conditions for the requalification of the labor force through technological change. It is here that the Marxian conception of socialism becomes more than a socio-political alternative and points toward fundamental civilizational change. But, where traditional Marxism assumed that workers would be guided by inherent and unambiguous interests in transforming technology, I will argue a contrary position, that *the institution of democratic control of industry is a condition for generating an interest in a new direction of technological progress*. In other words, democracy itself is a "productive force" of a new type, shaping the form of innovation in a future socialist society.[8] I would like to turn now to that argument.

A NON-DETERMINISTIC APPROACH

Marx's various scattered comments on the idea of socialism can be unified around two assumptions he routinely made, first that in industrial societies power depends on control of technology, and second, that the working class possesses an overriding class interest in the abolition of the division of mental and manual labor and the system of wage labor it implies. With these premises in mind, Marx argued that workers can rule industrial societies through the socialization of industry, a form of ownership in which the public character of modern forces of production is recognized through some sort of democratic control. The technological consequences of this arrangement follow from the imposition of new criteria of innovation responsive to the interests of the working class rather than to those of the capitalist class. Control of industry is thus more than control of society; it is also control of history. The class that decides on the course of industrial progress governs the future out of the present.

This brief sketch of Marx's views shows that he relies on the concept of interest to secure the connection between the democratic forms and the substantive technological content of his conception of socialism. The reference to interests gives an objective, imperative form to the process of the transition to socialism and suggests that, just as the interests of capitalists govern the cultural and technological development of capitalism, so the interests of workers would shape a new socialist civilization.

Since interests grow out of the relation of classes to the means of production, they can be objectively ascertained, independent of temporary ideological variations. On these terms, the chief interest of the capitalist class is the maintenance of the precondition for its existence as a class, namely, what I will call the *operational autonomy* required by its alienated administration of society's productive resources. This interest is secured in the system of property, which offers capitalists a great deal more freedom of action than had the leaders of work groups in earlier societies, freedom that can be used to increase their power and wealth. In particular, capitalists introduce technological innovations independently of the opinions and needs of workers and in function of the goal of maximizing control of the production process through technological design.[9]

The capitalist division of labor that results from this innovation process has a social impact through its cognitive effects on the working population. I will call these effects the *knowledge deficit,* defined with respect to the cultural level required for self-rule. The acquisition of knowledge generally follows the patterns established by the specialized breakdown of tasks that insures capitalist control of the labor process. Disqualification and increasingly narrow technical specialization confine workers to a dependent position in society, while casting capitalists and their managerial representatives in the leading role. The result is an enlarging gap between the level and type of culture required by most positions in the division of labor, and that required to make significant social decisions. Technological advancement in this system thus not only subordinates workers to capital, but disenfranchises them at the most fundamental level by removing all incentives for society to teach and for workers to learn enough knowledge about the social world to understand and participate in making the important social decisions that concern them.[10]

Capitalist technology has evolved under a system of class rule and therefore reflects the requirements of control from above. It is these requirements which generate a division of labor and a conception of technical progress incompatible with the full and democratic development of the individuality of the workers. A democratic power would have different requirements, tied to the civilizational project of creating an industrial society favorable to individual development.[11] It is important to note that Marx does not regard any of these points as politically contingent. When it comes to matters of control, it is not a question of what this or that individual or group prefers; essential determinants of class identity are involved. Capitalists *must* impose a division of labor which *only* workers can overcome. It is characteristic of Marxism to so condense a democratic process and a substantive goal through the identification of social essences.

The third dimension of Marxian socialism can now be reformulated more precisely in the following propositions, which express the interconnection between its political and technological aspects:

3.1. A socialist system would require the development of a qualitatively different division of labor compatible with the employment of highly educated and socially responsible workers.

3.2. Such a system would be oriented toward long-range patterns of technological development that would further the actualization of human potentialities at work.

The whole theory can now be summarized in what might be called Marx's *maximum thesis* on the transition, according to which:

> Workers' control of industrial society inevitably leads to the socialist transformation of the division of labor and technology.

This original Marxian position is subject to two very different kinds of attack today. First, it is claimed that the desirable social transformations Marx expected from socialism are in fact being made under capitalism in response to technological advance. Second, the deterministic reliance on class interest is rejected on the grounds that the self-understanding of social groups cannot be deduced from their place in the social structure but depends on a wide variety of factors. Let us examine these objections in turn.

1. Is the Marxian theory of capitalism refuted by the emergence of an "information" or "post-industrial" society, based on a radically new type of computer technology that privileges skills, in contrast with the deskilling associated with traditional mechanization? The initial determinist assessments of post-industrial society argued that computerization would raise the general skill level of the workforce, automate production and enlarge the range and power of human communication and collaboration at work, making

possible the general democratization of society.[12] But nearly a generation of experience with automation and computerization has changed the tone of social commentary.[13] An emerging consensus argues that automation is suspended politically between very different alternatives and nothing about computer technology per se can decide the issue. A brief glance at two recent studies of automation illustrates this point.

In her book *In the Age of the Smart Machine,* Shoshanna Zuboff argues that computers make possible two complementary transformations of the workplace.[14] On the one hand, they can be used to automate production, relieving human beings of physical effort. On the other hand, they can be used to "informate," a term Zuboff introduces to refer to the integration of workers and machines at a higher level of intellectual involvement, participation in decision-making and productivity. Informating is not an alternative to automation, but consists in further potentialities of computer technology management may or may not choose to realize in the course of automating. Zuboff offers persuasive evidence for the social ambiguities, or ambivalence, of computer technology.

Harley Shaiken concludes his book, *Work Transformed,* with similar reflections:

> It is ironic that computers and microelectronics should be used to create a more authoritarian workplace. They could just as easily be deployed to make jobs more creative and increase shop floor decision-making. Rather than pace workers, systems could be designed to provide them with more information about the production operation in general and their own jobs in particular. The technology could be used to bring the work under the more complete control of the people who do it rather than the other way around.[15]

Shaiken would seem to agree with Zuboff that, "Technological design embodies assumptions that can either invite or extinguish a human contribution."[16] And Zuboff offers numerous examples of the suppression of the democratic potential of informating technology by uncomprehending or frightened managers, confirming Shaiken's argument that automation has failed socially because of "the use of technology to extend managerial power."[17] Thus the issue for both Zuboff and Shaiken is less the impact of technical progress on the democratization of society than the reverse: the capacity of democratization to release suppressed potentialities of technical progress. But this was precisely the point Marx was getting at a century ago.

2. I would like to turn now to the far more damaging objection to Marx's deterministic conception of class interests. How strong, really, is the connection between working class power and the socialist society Marx expects

workers to create? Marx's original theory of working class interests assumed that capitalism could only satisfy workers' needs contingently, inadequately, and episodically. This biological basis for opposition to capitalism was linked to a sociological theory of alienation according to which the structural obstacles to self-actualization through work could only be overcome in a socialist society. Thus in this early formulation, workers' vital needs for food and shelter would be satisfied only where their human need for fulfilling work was also satisfied.

The later theory of "relative immiseration" recognized the possibility of a rising income floor without clearly explaining the impact of this reformulation on the motives and aims of socialist revolution. Once the impoverishment of the working class is defined as relative, socialism can no longer pretend to be the only adequate representation of workers' rational self-interest. A more sophisticated formulation is required which distinguishes between the wide range of interests that can be reasonably imputed to workers, many of which are as well satisfied under capitalism as under socialism, and the narrower subset of that range which is systematically obstructed by the existing technology and division of labor and which can therefore best be achieved through socialist transformation of the technological base.

Once this concession is made, the deterministic framework collapses. It can now be asked what would necessarily orient workers toward the new concepts of labor and welfare that Marx identifies with socialism. The interests of workers under socialism are ambiguous and may not guarantee an evolution toward a truly different form of civilization. The idea that there is a necessary connection between working class interests, the quality of work, and the direction of development of industrial society is not persuasive in its original deterministic formulation, but it can be reconceptualized more modestly as a theory about the ambivalence of the economic culture of capitalism.

The *facticity* of the worker under capitalism, as object of capitalist control, involves discontent over the nature of work.[18] This cultural structure can, but need not, inspire workers to make the sort of demands that would lead to a socialist transformation of the production system. In fact, there is a good deal of evidence from the existing Communist societies that controversial trade-offs are involved in the choice between working conditions and consumer goods. But no group other than workers can consistently support an end to the conditions of control from above, since such changes are subversive of every way of organizing the economy from outside the labor process itself.[19] This *minimum thesis* might be formulated to say that:

Workers' control is *uniquely compatible* with changes in technology and the division of labor in the interests of the systematic enhancement of human capacities.

This formulation of the idea of socialism requires the translation of Marx's theory of the transition out of the objectivistic language of "interest" into the more empirical language of culture. The concept of the *economic code*, referring to empirically identifiable cultural differences in economic values, can serve for this purpose. By the economic code, I mean the way in which individuals customarily perceive their own welfare, what they regard as economic goals, and what they consider to be legitimate or desirable economic means. The economic code governs the goods individuals seek to obtain as economic rewards, whether they expect these goods to be delivered publicly or as privately consumable commodities, which goods are signified as utilitarian, and which possess aesthetic, prestige, or other types of values, and so on. Every society also has characteristic economic codes concerning such things as workmanship, authority relations on the job, savings and leisure, the occupational expectations associated with various jobs, with sex and age, and so on.[20]

These cultural differences have major implications for the process of technological development. In any society, it is through encoded perceptions of the economy that the so-called economic "realities" take on their specific significance. In particular, culturally relative conceptions of means and ends and divergent interpretations of welfare govern innovation and the reception and adaptation of imported technology.[21] Technological futures are thus determined not only by technical factors but by economic culture as well.

With these considerations in mind, we can ignore Marx's maximum thesis and reformulate his theory of the transition in ideal-typical terms. For our purposes, it is unnecessary to prove that working class rule guarantees a socialist evolution of society. The interesting point is the *possibility* that workers in some socialist society might choose an original technological future corresponding in its main outlines with Marx's concept of the transition. The actual economic code governing workers' economic perceptions *may* become the basis for the adaptation of technology to socialist purposes where it approximates to the hypothetical "interests" Marx imputes to them. This contingent relationship between workers' economic culture and the preconditions for the transition to socialism renders the socialist future itself contingent, since under the assumptions of the minimum thesis socialism can only be created by workers' control and not, for example, by a suitably enlightened political dictatorship.

TECHNOLOGY AND THE DISTRIBUTION OF CULTURE

Marx's hypothetical construction of the interests of workers and his predictions about the future have been criticized and defended ad nauseam. Rather than continuing that rather fruitless debate, I intend to pursue the "minimalist" strategy outlined above by reformulating the concept of proletarian interests as the ideal type of a *socialist economic code*. This approach reveals the underlying economic logic of a new form of civilization that would not merely add a few desiderata to the current repertoire of "values," but which would have a truly different structure. On this basis, it will be possible to anticipate changes as fundamental as those which gave rise to citizenship through the abolition of estates, or the invention of childhood through a series of social changes culminating in the limitation of the labor market in the late nineteenth century. Marx's *Grundrisse* provides a basis for this approach which, with a certain amount of imaginative interpretation, can be substituted for the usual deterministic account.

In this text, the dynamic of technological change under socialism is guided by workers' "interest" in humane work settings, an interest which must be imposed on the inherited technological base through rationalization and innovation in the course of the transition. Marx believed that eventually workers would produce a whole new technology in which work would be "life's prime want" and not a burdensome obligation.[22] This goal would be achieved when labor

> is of a scientific and at the same time general character, not merely human exertion as a specifically harnessed natural force, but exertion as subject, which appears in the production process not in a merely natural, spontaneous form, but as an activity regulating all the forces of nature.[23]

The transition to this higher type of industrial society would involve a deep change in economic culture. Capitalist society, Marx argues, distributes wealth in the form of ever more varied commodities, but this commodity form is only a limited reflection of the actual enrichment of the human species represented by the growth of the needs and faculties of the consumers. It is this latter that is "real" wealth, the development of human attributes and capacities as dimensions of individual self-actualization, mediated by material goods to be sure, but not identical with them. Marx writes,

> In fact, however, when the limited bourgeois form is stripped away, what is wealth

other than the universality of individual needs, capacities, pleasures, productive forces, etc., created through universal exchange? The full development of human mastery over the forces of nature, those of so-called nature as well as of humanity's own nature? The absolute working-out of his creative potentialities, with no presupposition other than the previous historic development, which makes this totality of development, i.e., the development of all human powers as such the end in itself, not as measured on a *predetermined* yardstick? Where he does not reproduce himself in one specificity, but produces his totality? Strives not to remain something he has become, but is in the absolute movement of becoming?[24]

The extension of transport and communications is a clear example of Marx's new standard of wealth. Peasants confined mentally and physically to the small villages of their ancestors are "poor" by this standard, compared with modern individuals situated at the nexus of cosmopolitan interactions. Whether or not we share Marx's disdain for the countryside, the economic implications of his argument are clear. Instead of appearing in the form of goods external to the individual, wealth now appears as the developed powers of that individual. There is a sense then in which training and education, variety of experience and occupation become a higher type of consumer good. A socialist society values such an enlargement of human experience and individuality as an end in itself, without subordinating these forms of wealth to the pursuit of a profit on the sale of the material goods associated with their acquisition.

Why would this change in the social definition of wealth occur under socialism? Marx argues that the industrial economy not only produces a huge variety of commodities, but it also creates unique opportunities to apply the expanded powers of the individual productively.[25] These opportunities would be seized under socialism as economic motivations for the enlargement of human capacities. Once a society has been gripped by this dynamic relation of consumption to production – the pursuit of "real" wealth contributing to its production – radical changes in the economy will occur. Every activity that increases the skill and intelligence of the worker increases the value of labor power. Meanwhile, work itself becomes one important arena in which to develop the powers of the individuals.

But work remains work, however fulfilling. Thus even under socialism workers will strive to reduce the time spent working while simultaneously increasing their leisure, much of which would be used for learning. But the more workers employ their leisure to learn, the more productive their labor and consequently the shorter the workday. "The saving of labour time [is] equal to an increase of free time, i.e., time for the full development of the individual, which in turn reacts back upon the productive power of labour as itself the greatest productive power."[26] Socialist "interests" and the corresponding patterns of consumption develop the "wealth" of the individual personality and the productivity of labor in a self-reinforcing cycle.

This is truly a utopian conception, but one that is considerably more interesting than the utopias usually attributed to Marx. Here the economic circle is squared by the creation of an economic *perpetuum mobile* that feeds off the very resources it consumes. If in fact a primary leisure activity, pursued voluntarily and for its own sake, increased the value of labor, then it could be freely converted into an economic input. Consumption and investment would approach identity in the domain of human resources and one of the chief costs of the economy, education and training, would also be a benefit.[27]

At the civilizational level, the key difference between all present day industrial societies, including Communist ones, and the socialist model outlined above is the role of knowledge and skill in the economy. Under socialism the "subjective" forces of production – human skill and intelligence – formerly wasted by industrial societies, could be applied to increasing productivity according to the dynamic of requalification projected in the *Grundrisse*. The socialist labor process emerges as a synergism of the demand for skilled labor and the growth of human powers through work. The higher level of knowledge and skill characterizing this labor process would make new efficiencies possible, motivate the transformation of technology, and reconcile broader participation with the technical requirements of an industrial society.

So formulated, the theory of the transition to socialism is an attempt to show the *possibility* of an alternative type of industrial society, one based on a much fuller development of individual capacities than in capitalism. By "possibility" in the sense of the term used here, is meant not merely technical feasibility, but the integration of technical and economic codes in a configuration corresponding to a new type of civilization in which the very pursuit of economic efficiency would raise the cultural capacities of individuals to the level of the social responsibilities they would have to fulfill, in which there would be no necessary trade-off between the pursuit of economic well-being and democratic participation.

THE TRANSITION TO SOCIALISM

Can we bring this utopian conception down to earth to inform our speculations about the future? I will argue that the effective initiation of a transition to socialism has three specific institutional correlates in the economic domain, corresponding to the three dimensions of socialism discussed at the beginning of this article. While these are not sufficient conditions for achieving a socialist reorganization of society, they do appear to be necessary conditions; and the discussion of them offers a useful way of concretizing the model of democratic socialism proposed here.

First, under the assumptions introduced above, social ownership must extend not just to machines, buildings, and land, but also to the monopolized cultural capital required for the management of industry. But the *socialization of cultural capital* cannot be accomplished at the stroke of a pen; it takes a long term educational effort on an unprecedented scale.[28]

Furthermore, if workers are to assume greater responsibility for their enterprises, the systematic redistribution of cultural capital must begin in the very earliest stages of the transition to socialism and must aim at raising the general cultural qualifications of the population. Thus it is not enough that the society favor study; narrow, merely technical training for existing positions in the division of labor must be replaced by much broader and continuing preparation for social participation. Workers must be in a position to enlarge their access to the kinds of knowledge required to perpetuate and increase their power in society.

The democratic distribution of culture would accomplish two fundamental objectives:

1. to supply the volume of intellectual resources required to take advantage of technological options that rely more heavily than does capitalist technology on inputs of skill and intelligence;

2. to qualify the entire labor force, and not just a small elite, to participate effectively in management and politics.

This strategy would seem to involve wasteful investments in human resources. But this objection is rooted in the culturally relative distinction between *investment* and *welfare* in the capitalist economic code. We tend ethnocentrically to consider these as unambiguous categories because for us the goals of production are signified in terms of capitalist concepts of wealth, that is to say, primarily in the form of material goods available for private consumption. By contrast, education is an investment rather than a major positive component of personal welfare. The scarcity of knowledge and skill is a direct result of this code, which regulates supply by market demand and which rewards deskilling with a share of the savings realized by the replacement of skilled with unskilled labor.

Following Marx's argument in the *Grundrisse*, we could construct an ideal type of a socialist economic code in which, on the contrary, educational activities that capitalist society places within the category of investment and evaluates in terms of efficiency in the production of material goods would be placed in the category of consumption and evaluated as contributions to individual welfare. While this is a promising beginning for a theory of socialist economic culture, Marx's account of how such a culture would come to prevail over the capitalist one is unacceptable.

The shift would occur, according to the *Grundrisse,* because technological progress would render the capitalist method of distributing culture obsolete and would usher in a new age in which the acquisition of knowledge and skill becomes an end, an activity in which workers will voluntarily engage independent of personal economic return. Industrial advance, beyond a certain point, would yield a new social definition of welfare as self-actualization. This argument is a clear instance of Marx's tendency toward technological determinism, and as such unpersuasive.

The collapse of the deterministic position forces us to answer a type of objection Marx never had to face. Socialist educational strategy sounds remarkably idealistic in the bad sense of the term. Indeed, today education is usually treated as a moral value where it is not merely an economic investment, suggesting that democratic socialism would be committed to the voluntaristic imposition of educational goals. On these terms, the socialist hypothesis that education would be pursued for its own sake is not much more plausible than the hope in a general commitment to moral self-improvement. Is there a way to retrieve Marx's notion of educational consumption from these consequences of the breakdown of his determinism?

If the pursuit of education is to be a consumer good under socialism, it must serve importantly in a practical context of daily life. Educational consumption must have some purpose more compelling than the sheer enjoyment of learning.[29] The elements of a solution are present in the democratic dimension of socialism, but have never been applied to the problem before. The local politics of participatory self-management in workplaces and in communites would offer a far more effective scene for the application of broadened cultural capacities than the inherited labor process. Thus, to a considerable extent, the consumer value of education would be realized at first in relation to public functions such as management and community affairs. There educational acquisitions would "pay off" for the individuals, if not economically, at least in terms of increased respect and influence and overall outcomes more in conformity with their needs and interests.

The scope of education would broaden accordingly, and in this the acquisition of knowledge and skill would no longer appear as subtractions from individual welfare but as components of it. The educational activities of the population would be *uncoupled* from the immediate economic needs of the society and from the investment strategies of individuals; education would become the driving force in social and technological change. Industrial society would *bootstrap* from a condition characterized by the knowledge deficit to one in which individuals would generally possess the cultural qualifications corresponding to their social responsibilities.

With the growing success of the transition, the initial "overinvestment" in education would make possible a leap to a higher level of productivity, based on new technologies adapted to operation by a highly educated workforce. Not technology but politics must lead the process of change, with technological progress an outcome rather than a cause of the establishment of new socialist social relations. Thus once Marxist determinism is abandoned, *democracy becomes an economic and technological requirement of the transition to socialism.*

Although such an educational dynamic contrasts sharply with our expectations in the advanced capitalist world today, it has partial precedents in the modernization of Japan and the Soviet Union.[30] To dismiss it as unlikely in any particular contemporary context misses the point that it defines a possible trajectory of development toward a new form of industrial civilization in which social responsibility is much more widely distributed than in today's world. Those who would seek an easier path to such an outcome must explain how it will be possible to achieve radical democratization in a technological society on the basis of the low levels of culture and intellectual qualification determined by the existing division of labor.

The emphasis on education is thus in harmony with a *second* requirement of the transition, the *democratization* of society through the establishment of formal institutions workers can use to intervene in the activities of those who exercise power and authority in their name. This question is inseparable from the problem of expertise because, as I have argued above, the subordination of workers is not merely a political matter but is rooted in the division of labor introduced by capitalism. The transition aims to create a new state of affairs in which *the recomposition of formerly divided mental and manual labor reduces the operational autonomy of leadership and reincorporates the alienated functions of management back into the collective laborer.*

This approach, which Marx and Engels never adequately theorized, might be explained by borrowing Habermas's distinction between forms of rationality. Under capitalism communicative and technical rationality become competing bases of social power, their opposition growing out of a social structure which allocates power to those performing a special technical function, the leadership of the collective laborer. But these two forms of rationality are not *essentially* in conflict. It is only in the peculiar framework of modern capitalism (and its Communist imitators) that the technical sphere offers a basis for advancing claims to power in competition with claims made on the basis of the communicatively rational appeal to common standards and interests. A socialist society would therefore attempt to modify that framework in order to reduce the relative weight of technically based claims to power in favor of communicatively based claims.

Just how far can these technically based claims be rejected in the early phases of the transition without disrupting production? This is a critical question, but Marx never addressed it, no doubt because he could not respond a priori. But unfortunately those in a position to offer a response at a later stage did not even know how to pose the question, much less answer it. Thus the lack of an early industrial movement for workers' control on the basis of which Marx might have projected the first stages of economic democratization under socialism turns out to be an historical accident heavy with consequences.

With no guidance on the question of workers' control from Marx, later Marxists treated capitalism exclusively as a form of ownership, and its management system was identified with the general technical requirements of industrial production. In this context, the Russian and European council movements of 1905 and 1917–1919 looked like just another form of political mobilization and were praised or condemned as such without any appreciation for their profound social significance. Even democratic socialists believed for the most part that authoritarian control from above was an economic necessity. But such control could only be reestablished by once again disenfranchising workers in the workplace, in effect ending the revolution. Workers thereby lost any chance of initiating a real transition to socialism.

These are very much live issues today, as the Soviet Union, Poland, and Hungary struggle to reconstruct their Communist systems in a more democratic and economically efficient form. The dismantling of bureaucratic dictatorship and its clumsy planning machinery requires decentralization and greater reliance on markets, either through a capitalist option such as denationalization of government owned enterprises, or a democratic socialist option such as their autonomization under the control of managements chosen by the workers. There is a considerable literature on the socialist solution, derived for the most part from the Yugoslavian experience with self-management.[31]

Self-management is not, however, without its problems, many of them evident in the Yugoslavian case. One school of thought argues that for self-management to succeed, formal democratization of the selection of managers is a necessary but not a sufficient condition. What might be called *deep democratization* of technically based power structures is required as well. This second requirement results from the fact that the socialist workforce must continue to rely on the authority of highly trained professional and managerial personnel for a prolonged period, no doubt measured in generations rather than years or decades as some have hoped. Their authority must be made to contribute to the transitional process through the way in which it is accommodated to the gradual enlargement of workers' initiative and control. The real problem of democratization lies here.[32]

Deep democratization implies significant changes in the structure and knowledge base of the various technical specializations underlying technical and administrative work to the extent that these serve as bases for advancing unwarranted claims to power. The rules and roles governing the exercise of authority must be altered to promote greater autonomy on the part of the "client" subject to the authority. These rules and roles concern such things as the method of training and selection of managers, the social context of management, and the procedures and criteria of innovation.

Other agent-client relations outside the domain of business are modeled on the capitalist project of maximizing control from above. Thus to the extent that socialism prevails in industry, changes can be anticipated throughout society, in government administration, the position of science and science based technical systems in society, the organization of medical practice and the "sick role," the craft rules governing mass media production, teaching, and so on.

How realistic is this strategy for recomposing the unity of the collective laborer? The chief obstacle would appear to be the opposition of technically qualified personnel to such a shift in allegiances and work conditions. But historical experience is not unambiguous and suggests a possible trajectory of change. Although it is not a fact given sufficient weight and notice, modern revolutions are characterized by sporadic "fraternization" between workers and sympathetic members of the bureaucracies and corporate administrations. Often in the enthusiasm of the moment quite elaborate plans for reform of management and government agencies are proposed, as was actually the case in the French May Events of 1968.[33] These experiences indicate that the idea of reorganizing the collective laborer in new ways is not merely idle speculation but resonates with historical experience.

To go beyond such anecdotal evidence, a theory of the middle strata is required.[34] But such theories generally founder on the attempt to explain their object in terms of conventional concepts of class struggle and alliance, which generally presuppose that the classes have clearly defined interests independent of the political relations established in the alliance. But it is precisely the breakdown of such prior definition which seems to characterize the behavior of the middle strata in a crisis. The position of these groups is better understood by abandoning traditional class theory and returning to simpler, more empirical concepts. Then it is evident that neither the capitalist nor Communist middle strata can be satisfactorily conceived as a class because they owe their existence to their *place in organization* rather than to an economic function or role.

These groups are composed of individuals who have been hired, after meeting requirements such as educational credentials, to carry out an action based on specific technical codes. These codes embody basic social and technical assumptions about how to design and operate technical systems. Thus unlike the other classes of modern society, which arise from an "organic" economic process, the members of the bureaucracies acquire their class identity through a process of selection, rooted in a relationship to technology which qualifies them to represent others. This is the origin of the ideology of service which we associate with "professionalism" in the West. In professionalist ideology the middle strata are the "agents" of "client" individuals and groups in whose interests they act, and for whom they perform services these latter cannot perform for themselves.

This explains why, when the "clients" rise in struggle, the legitimacy of the bureaucracy's selection and conception of service is questioned on a global scale. Its claim to represent the public interest is visibly shaken and its technocratic self-image enters into crisis. The people can then appear "in person" as the source of a legitimacy granted hitherto by capitalist or Communist elites in their name. The people are a recourse and an ally through which the middle strata can be reconstituted and their "selection" reconfirmed under a different hegemony for different social purposes. The culmination of such a reconstitution of the bureaucracies would be the elaboration of new practices and technical codes for each field of work aimed at reducing the alienation (in system terms, the "operational autonomy") of professional leadership.[35]

In sum, where popular initiatives provoke splits in the ranks of the bureaucracy, its crisis can be understood as an attempt to reconstitute the unity of the collective laborer through the innovation of new practices and codes. The bureaucracy is no longer an interest in its own right engaged in maximizing its operational autonomy at the expense of workers, but becomes instead a scene of struggle in which popular interests are actively represented. These interests may inspire a fundamental restructuring of technology and society, initiating a process of civilizational change.

A *third* requirement of the transition concerns the direction and organization of *innovation* under socialism. During the transition, economic and technological development would proceed in a very different environment from Western capitalism. Skilled labor would be far more abundant than it has been at corresponding stages in the Western process of modernization, limited primarily by its social cost once private costs had been reduced or disappeared in the accumulation of educational resources. Under these conditions, *highly qualified human resources would not be in short supply but would be widely available as a nearly "free" good* on which the economy could draw at will.

In addition, the democratization of management would have profound effects on patterns of innovation, altering the technical code in terms of which progress is defined. The capitalist code, adjusted to the need to maximize profit and control the work force, would be replaced by a different code, still concerned with efficiency but taking into account a wider range of variables. As Carol Gould writes, in comparison with capitalist managers,

> members of a worker self-managed firm would be prone to be more sensitive to the impact that the use of given technologies would have on their conditions of work and the quality of work life. They might well also be responsive to issues of consumer need and environmental effect, since they are themselves also consumers and residents of the local area.[36]

An economy developing under these conditions might reach efficient solutions to technical problems in new ways, moving generally toward an alternative type of industrialization. In some cases skills might be applied in relation to technologies that it would be economically irrational to employ in a capitalist society because of the very different organization of human resources under capitalism. In other cases, a different approach to work conditions and the environment might result in innovations that would not have occurred or succeeded under capitalism. Different patterns of consumption and different leisure pursuits would occupy the labor force of such a society, and in it the political process might take on a qualitatively different character. This would be, in short, a socialist system of production, one in which technological change would be governed by new principles.

There are two commonplace objections to this view which need to be addressed briefly here, although they are really the subject of a historically and sociologically concrete discussion of innovation. The first of these objections is the familiar argument that planning necessarily suppresses the individual freedom and initiative required for innovation. Indeed the standard image of the innovator in capitalist society conforms to the romantic myth of individual genius, not uncommonly at odds with ignorant bureaucrats.[37] Interestingly, innovation continues to occur in certain large capitalist corporations which have directly confronted the problem of stifled creativity by opening spaces for innovation within their boundaries through radical departures from classical organizational models. Small teams are allowed to work in an almost parasitic relation to the corporation, drawing on its resources for an unusually autonomous activity of research and development.[38] Much can be learned from such examples. The transition to socialism requires the working out of comparable organizational innovations as the basis for the technological changes it must make in the creation of a new type of industrial civilization.

The second objection argues that technological development is essentially determined by technical considerations, largely independently of the social environment in which it occurs. Socialism, if it is possible at all, would therefore have to prove itself compatible with technology rather than the reverse. Usually this argument culminates in the demonstration that all technological societies must come to terms with the "imperative" requirement of authoritarian management. This is technological determinism, a view which has always been more popular in development theory than among historians of technology. I address this problem at length in my forthcoming book, *The Critical Theory of Technology,* but can do little more here than note the existence of an important literature in support of the view that technological development is a dependent variable in the social system.[39] This view suggests the possibility of anticipating the broad social determinants of technological progress in a socialist society as I have attempted here.

This article began by suggesting that democratic socialism involves a process of civilizational change more complex than anything we would normally consider under the heading of politics. These concluding reflections attempt to identify possible starting points for such a process. The result is not a utopian description of a perfect society, but rather a hypothetical direction of social development which would lead to a new form of civilization through democratic changes at many levels: political, economic, cultural, and ultimately technological.

San Diego State University

NOTES

* This article is drawn from my forthcoming book, *The Critical Theory of Technology*. I would like to take this opportunity to thank Frederick Fleron, Jr., Gerald Doppelt and Langdon Winner for criticism and encouragement in the development of the ideas expressed here.

[1] For a wide ranging survey of the contemporary discussion of socialism and democracy, see Frank Cunningham, *Democratic Theory and Socialism* (Cambridge: Cambridge University Press, 1987). See also Carol Gould, *Rethinking Democracy* (New York: Cambridge University Press, 1988).

[2] "What I did that was new was to prove: (1) that the existence of classes is only bound up with particular, historical phases in the development of production, (2) that the class struggle necessarily leads to the dictatorship of the proletariat, (3) that this dictatorship itself only constitutes the transition to the abolition of all classes and to a classless society." From a letter of Marx to Weydemeyer dated March 5, 1852. In V.I. Lenin, *Selected Works* (New York: International Publishers, 1967), volume 2, p. 291.

[3] Marx wrote, "Communism is for us not a *state of affairs* still to be established, not an *ideal* to which reality [will] have to adjust. We call communism the *real* movement which abolishes the present state of affairs. The conditions of this movement result from premises now in existence." Loyd D. Easton and Kurt H. Guddat, eds., *Writings of the Young Marx on Philosophy and Society* (Garden City, N.Y.: Doubleday, 1967), p. 426. Paul Sweezey denies there is such a "law" in Marxism, writing that "The assumption, more often implied than spelled out, is that once socialism...has been firmly established, *its own inner dynamic will automatically propel it forward on the next leg of the journey to communism....* No one, however, has succeeded in explaining what the 'law of motion' of socialism...is supposed to be." Paul Sweezey and Charles Bettelheim, *On the Transition to Socialism* (New York: Monthly Review Press, 1971), p. 125. I suggest that the idea of "possible trajectories of development" is present in Marx and that he may have mistaken it for "laws."

[4] For more on this concept, see Herbert Marcuse, *One-Dimensional Man* (Boston: Beacon, 1964), pp. 219 ff. For an account of Marcuse's theory of potentiality, see my "The Bias of Technology," in R. Pippin, A. Feenberg, and C. Webel, eds., *Marcuse: Critical Theory and the Promise of Utopia* (Amherst, Mass.: Bergin and Garvey, 1988).

[5] The two most important texts for understanding the Marxian theory of the transition are, "The Critique of the Gotha Program" and "The Civil War in France." The relevant passages are published in Robert Tucker, ed., *The Marx-Engels Reader* (New York: Norton, 1972), pp. 383–398, and pp. 526–576.

[6] Jürgen Habermas, *Toward a Rational Society* (Boston: Beacon, 1970), p. 87. I will assume here that technology may be considered socially determined in some significant dimensions. My arguments for this position are offered in "The Bias of Technology," in Pippin, Feenberg, and Webel (note 5, above), and in "The Ambivalence of Technology," *Sociological Perspectives* (Fall, 1989).
[7] For examples, see Ernesto Laclau and Chantal Mouffe, *Hegemony and Socialist Strategy* (New York: Verso, 1985); Carl Boggs, *Social Movements and Political Power* (Philadelphia: Temple University Press, 1986); Jean Cohen, *Class and Civil Society* (Amherst: University of Massachusetts Press, 1982).
[8] The economic significance of democracy is discussed in Pierre Dokes and Bernard Rosier, *L'Histoire ambigue* (Paris: PUF, 1988), pp. 291–294.
[9] The basis for this approach to Marx's work is Part 4 of Volume **1** of *Capital* (New York: Modern Library, 1906). For recent accounts, see Harry Braverman, *Labor and Monopoly Capital* (New York: Monthly Review Press, 1974); Paul Thompson, *The Nature of Work* (London: Macmillan, 1983); and Ali Rattansi, *Marx and the Division of Labour* (London: Macmillan, 1982).
[10] Marx, *Capital,* volume 1, p. 461. For a discussion of this problem, see my "Transition or Convergence: Communism and the Paradox of Development," in F. Fleron, Jr., ed., *Technology and Communist Culture* (New York: Praeger, 1977).
[11] On the status of self-actualization as a socialist value, see Gould, *Rethinking Democracy* (note 1, above), chapter 1.
[12] John Diebold, *Automation* (New York: Van Nostrand, 1962), p. 162.
[13] For examples, see Walter Buckingham, *Automation* (New York: Harper & Row, 1961), pp. 96 ff; and David Noble, *Forces of Production* (New York: Oxford University Press, 1984).
[14] Shoshana Zuboff, *In the Age of the Smart Machine* (New York: Basic Books, 1988).
[15] Harley Shaiken, *Work Transformed* (Lexington, Mass.: Heath, 1984), p. 267.
[16] Zuboff, *Age of the Smart Machine* (note 14, above), p. 182.
[17] Shaiken, *Work Transformed,* p. 268.
[18] See *Work in America: Report of Special Task Force to the Secretary of Health, Education and Welfare* (Cambridge, Mass.: MIT Press, 1973).
[19] For a recent attempt to understand this facticity of classes and its relation to ideology, see Goran Therborn, *The Ideology of Power and the Power of Ideology* (London: Verso, 1980).
[20] For a far reaching critique of the concept of interest in Marxism, see Laclau and Mouffe, *Hegemony and Socialist Strategy* (note 8, above). For a discussion of economic codes, see Jean Baudrillard, *Pour une Critique de l'economie politique du signe* (Paris: Gallimard, 1972).
[21] A literature has grown up in the field of international development to attempt to account for these differences, which complicate technology transfers. See Edward Spicer, ed., *Human Problems in Technological Change* (New York: Russell Sage, 1952). For an example of such an approach to Communist societies, see Fleron, *Technology and Communist Culture,* note 11, above.
[22] Marx, "Critique of the Gotha Program," p. 388.
[23] Marx, *Grundrisse* (Baltimore: Penguin, 1973), p. 612.
[24] *Ibid.,* p. 488.

[25] Marx, *Capital*, volume 1, pp. 533–534.
[26] Marx, *Grundrisse*, p. 711–712.
[27] Educational programs that required full-time attendance would still represent significant costs for the individuals; but part-time adult education, pursued as a leisure activity, would fall in a different category and might make a large free contribution to the economy. For the distinction between these different costs, see Gary Becker, *Human Capital* (Chicago: University of Chicago Press, 1975), pp. 194–195.
[28] On the concept of cultural capital, see Alvin Gouldner, *The Future of the Intellectuals and the Rise of the New Class* (New York: Seabury, 1979). See Rudolph Bahro, *The Alternative in Eastern Europe*, London: New Left Books, 1978): "Marx and Engels...sought to intervene actively in the abolition of traditional labour, with a process that I would today describe with deliberate political intent as a cultural revolution, since they believed – *at the time* somewhat optimistically – that the general level of productivity was already sufficient to set free sufficient 'disposable time' for the development of the general abilities of all people, by the participation *of all* in necessary labour" (p. 278).
[29] In modern economic theory, educational consumption is treated exclusively in terms of its potential for enhancing the value of future consumption of goods through the refinement of appreciation. This is an excessively narrow basis on which to conceptualize the leap in cultural level required by a radical civilizational change, and, to be fair, the theory was certainly not introduced for that purpose. See Becker, *Human Capital* (note 20, above), p. 69.
[30] For more on the question of early Soviet education, see Kendall Bailes, *Technology and Society under Lenin and Stalin* (Princeton, N.J.: Princeton University Press, 1978).
[31] There is a large literature on the concept of democratic management. See, for examples, Paul Blumberg, *Industrial Democracy* (New York: Schocken, 1976); and Pierre Rosanvallon, *L'Age de l'autogestion* (Paris: Seuil, 1976). For an evaluation of Yugoslavia in Marxist terms by a theoretician of self-management, see Mihailo Markovic, *From Affluence to Praxis* (Ann Arbor: University of Michigan Press, 1974), chapter 4. A skeptical alternative view is presented by Ellen Comisso, *Workers' Control under Plan and Management* (New Haven, Conn.: Yale University Press, 1979). For a recent philosophical defense and justification of self-managing socialism, see Gould, *Rethinking Democracy* (note 1, above), chapters 4 and 9.
[32] Arguments for this conclusion are offered in articles by Gorz, Maccio, and Il Manifesto in Andre Gorz, ed., *The Division of Labor* (Sussex: Harvester, 1978).
[33] See Andrew Feenberg, "Remembering the May Events," *Theory and Society* (July, 1978).
[34] For a collection surveying the main themes of the debate on the class status of the middle strata, see Pat Walker, ed., *Between Labor and Capital* (Boston: South End Press, 1979).
[35] During the Cultural Revolution, attempts were made to study China on these terms. See, for example, Rensselaer Lee's comments: "The function of 'politics' in Communist China is largely to distribute opportunities of generating technological and cultural change. This redistribution occurs at the expense of professional elites and results in a close integration of change-producing actions with participations in

labor"; in "The Politics of Technology in Communist China," *Comparative Politics* 5 (1973): 323.

[36] Gould, *Rethinking Democracy* (note 1, above), p. 277.

[37] For discussions of the actual problems of innovation in Communist societies, see R.V. Burks, "Technology and Political Change," in C. Johnson, ed., *Change in Communist Systems* (Stanford, Calif.: Stanford University Press, 1970); and Rensselaer W. Lee, III, "Mass Innovation and Communist Culture: The Soviet and Chinese Cases," in Fleron, *Technology and Communist Culture* (note 10, above). For a classic discussion of the wide variety of contexts in which innovation has occured historically, see John Jewkes, David Sawers, Richard Stillerman, *The Sources of Invention* (New York: St. Martin's, 1959).

[38] For an example, see Tracy Kidder, *The Soul of a New Machine* (New York: Little, Brown, 1981).

[39] See, for example, Arnold Pacey, *The Culture of Technology* (Cambridge, Mass.: MIT Press, 1983).

STEVEN L. GOLDMAN

PHILOSOPHY, ENGINEERING, AND WESTERN CULTURE

Engineering has long been treated with condescension in Western culture and this continues today even among those intellectuals who have discovered the cultural significance of science and, very recently, of technology. In a complementary essay to this one, I describe technology as a social process to which engineering contributes, but which is driven by institution-specific executive decisions that apply technical knowledge selectively to the accomplishment of managerial agendas.[1] The practice of engineering, I argue there, is captive to social determinants of this process such that the definition of engineering problems, the determination of the means to be used in solving them, and the identification of what will count as solutions, all derive from the institutional context of engineering's practice, not from the knowledge engineers possess, and certainly not from Nature.

In this essay, I wish to explore engineering's *theoria*, that is, its characteristic world-view and rationality. Both of these are quite different from the world-view and the rationality of the physical sciences, whether traditionally conceived as bodies of objective knowledge or, on the so-called post-positivist view, as bodies of socially-constructed knowledge. In order to discuss engineering's *theoria*, however, it is first necessary to confront two obstacles to achieving an understanding of engineering more adequate to an appreciation of the intellectual problems that it poses. The first obstacle is a persistent Western cultural prejudice that subordinates engineering to science. This interpretation of engineering practice as "merely" applied science denies the existence of an independent theory of engineering by supposing that the principles of engineering reasoning are just those of physical science. The second obstacle derives from the fact that engineering serves processes of technology-related social change. As a consequence, it is caught up in the struggles for social, political, and economic power that are inevitable accompaniments of such change.

The claim that engineering is applied science rests on the assumption that physical science and engineering share a common understanding of the world and its properties, an understanding based on a shared body of knowledge

Paul T. Durbin (ed.), Broad and Narrow Interpretations of Philosophy of Technology, 125–152.

generated primarily by scientists and always by "the scientific method." A presupposition of this assumption is that science and engineering take the same world as their object. Neither of these seems to me to be true. Scientific knowledge and engineering knowledge are two fundamentally different kinds of knowledge, and, bizarre though it may sound at first hearing, they have different worlds as their objects.

If this assumption and its presupposition *were* true, that is, if science and engineering did share a common understanding of the same world, then engineering would pose no epistemological or metaphysical problems that were not already posed by the sciences. Engineering's ontology would be the ontology of the physical sciences. Its theoretical knowledge would be provided for it by scientists or, in a minority of cases, by engineers using the scientific method and thus acting as proxy scientists. Engineering practice would be the empirical qualification, as required by the local circumstances of a particular application, of universal theoretical knowledge, complicated in detail perhaps, but hardly of compelling intellectual interest. There would, in short, be no need for a philosophy of engineering analogous to the philosophy of science.

To be sure, even on this view their common world would be experienced differently by engineers and by scientists, because to make their applications "work," engineers have to wrestle with the refractoriness of the particular in its particularity. For scientists, by contrast, the particular appears as an instantiation of a universal. As a consequence, the particular is so much better behaved in the world of the scientist than in the world of the engineer that, as it can be deduced from the universal, its particularity can be ignored. This differential experience even of the world they are supposed to share has not, however, led philosophers of science to propose that engineering has an understanding of its own, reflecting its characteristic experience of a world more resistant to "pure thought" than the world experienced by the scientist.

One reason why this has not happened is that for the overwhelming majority of Western intellectuals, from Plato to the present, reasoning is the means by which we discover reality and determine truth, and the form (though not necessarily the content) of reasoning is expected to be context-free. That is, it is expected that reasoning will take the same form in every application: in philosophy, in mathematics, in the physical sciences, and, if they are to be rational enterprises, in the social sciences and in engineering as well. Furthermore, reasoning is supposed to be objective, in the sense of impersonal and free of willfulness. In addition, there is a strong Western prejudice in favor of universal and necessary inference, so that deduction and not induction is identified as *essential* reasoning, the form of reasoning in which reality and truth are apprehended least equivocally. Finally, the Western intellectual tradition displays a clear preference for understanding over doing, for contemplation over operation, for theory over experiment.[2]

As a matter of (socially constructed?) fact, reasoning does take this one universal, necessary form in mainstream Western philosophy, mathematics, and physical science. This is reflected in the obsessive quest for certainty in Western thought – a quest Erasmus called "the comedy of the higher lunacy"[3] – and in corollary obsessions with deductive arguments and with the need to overcome probabilism and refute skepticism. But reasoning does not take this same form in engineering. As a matter of principle, solutions to engineering problems have a particular, arbitrary, and contingent character. They cannot be deduced from the terms in which those problems are defined, even taken together with their boundary conditions and the mathematical models engineers employ. In addition, willfulness is inseparable from the definition of engineering problems and solutions to them; and engineering is, of course, fundamentally experimental and concerned primarily with doing, valuing theory as a means to operational ends. Both engineering problems and their solutions are inescapably context-dependent for their very definition.

The relationship of engineering to science bears some similarity to the relationship between mathematics and physics. Ostensibly, the mathematician and the physicist use the same mathematics. An observer sees the same symbols being used by both, sees the same rules used for their manipulation, and might conclude that physics and mathematics constitute a continuous body of knowledge. But the appearances are deceiving. Mathematics as used by physicists is fundamentally different from the mathematics that is the object of the professional mathematician. It is almost as if two communities used the same words in their languages, which had overlapping, but non-identical syntactic rules and semantic properties. Physics appropriates to itself certain products of mathematical research and, in the process of applying them to physics problems, transforms them. As a matter of principle, a mathematician could not step into a partially-worked-through problem in mathematical physics and continue its solution as a problem in mathematics, as, for example, a person could pick up a partially completed crossword puzzle and, working backward by reviewing the words already entered and forward over the empty spaces, complete it. At every step of their reasoning, physicists impose on the mathematical apparatus employed by them constraints deriving from the characteristics of the physics problem whose solution is their object.

These constraints range from limits, totally arbitrary from the perspective of the mathematician, set on integrals or on the ranges of functions in order to make operating with them more convenient, and the setting of boundary conditions and the elimination of infinities by mathematically arbitrary transformations, to relaxed notions of proof, completeness, existence, or continuity – again, relaxed from the perspective of the professional mathematician. To work on a physics problem, as opposed to contributing to such work, a mathematician would have to allow his or her reasoning to be shaped by the physics of that problem even as regards the use of mathematical techniques in solving it. Even at its most intensively mathematics-dependent, physics remains physics, not a branch of mathematics, not even of applied mathematics. Rather, mathematics is applied in physics, on physicists' terms, to problems defined by physicists, who also define when those problems are to be considered solved.

A similar relationship exists between science and engineering. Engineers do not simply take up the theories, models, or results of scientific research and use them to solve engineering problems any more than physicists simply take up the results of mathematics research and solve physics problems by "plugging in" mathematical functions. Engineers generate the knowledge that they need, to solve the problems they define, in terms they assimilate. To do so, they selectively appropriate scientific knowledge, in the process transforming it into engineering knowledge.

Engineering knowledge may thus be quite different from scientific knowledge, as mathematical physics is quite different from mathematics. Nor can it be taken for granted that scientific knowledge and engineering knowledge, though superficially different, essentially overlap. The cognitive character of engineering reasoning is informed by a distinctive intentionality that is embodied in the problems that engineers identify and in the approaches they adopt to solve them. By this I mean that engineers' thinking is significantly shaped by the objects about which they think as engineers, objects that range from the artifacts they handle and the physical processes they employ or can order to be employed, to their conceptualization of a given problem and of techniques relevant to solving it.

The same is true of physics and of mathematics, of course. That is, each of those disciplines, too, has a characteristic mode of thinking that is embodied in their respective conceptualizations of what their problems are, and of how solutions to them are to be achieved. As a consequence, philosophy of science is not obviated by philosophy of mathematics and neither should philosophy of engineering be obviated by philosophy of science.

Indeed, even under the rubric "philosophy of science," distinctions are made by philosophers between philosophy of the physical sciences, philosophy of the life sciences, and philosophy of the social sciences. In each case, philosophers have focused on the nature of the reasoning process in the discipline studied, the methodology employed that makes it "truly" scientific, and thus truly rational, and on the epistemological and metaphysical issues that need to be resolved in order for its rationality to be established to the satisfaction of the philosophical community. The nature of engineering reasoning calls for comparable consideration if engineering is to be understood, and not just for the sake of completeness.

Engineering poses very different, and much more complex, issues for philosophy from those posed by philosophy of mathematics and philosophy of the sciences. This is because the objects of engineering reasoning are far more complex than the objects of scientific reasoning; the former, unlike the latter, never lose their particularity and are *explicitly* inseparable from the intentional, contingent, willful, and value-laden contexts of their formulation. Engineering reasoning cannot be discovered from the logic of engineering practice which centers not on the application of a generic solution methodology, but on the explicitly value-laden and highly particular process through which engineering problems and solutions are specified. These specifications, in turn, are manifestly arbitrary. The distinctiveness of engineering *vis-à-vis* science derives from this specification process, which implicates the design dimension of engineering problem-solving (as I will discuss below). Design mandates a form of reasoning about engineering problems that is fundamentally incompatible with the universal, context-free conception of rationality traditionally promoted by philosophers and embedded in modern mathematics and physics. This is – it should be noted – not intended to imply either that rationality, as traditionally conceived, in principle *is* universal and context-free, or that its actual employment by mathematicians and scientists is universal and context-free; only that reasoning in modern science has been, and continues to be, conceived of by scientists as in principle universal and context-free; that is, free of historical, socio-cultural, and personal determinants of its logic and its objects.

The problems addressed by the sciences are, as revealed by the professional rhetoric of their practitioners, supposed by scientists to be given to them by "Nature." That is, scientific problems, in the first instance, are conceived of as addressing features of an independently existing reality. Within the practice of science, scientific problems are commonly conceived of as discovered; they are not arbitrary human inventions. Furthermore, because Nature is assumed by scientists to be the *uniform* ground of their self-given problems, it is possible to suppose that a single form of reasoning, the characteristically "scientific" form, will suffice to respond to all scientific problems. This supposition anchors the myth of a unique scientific method. Finally, solutions to scientific problems are also supposed by scientists to be given to them and on the same terms as the problems; that is, they, too, are independent of the reasoning process and of the personal, social, and historical circumstances of the reasoner (except incidentally and accidentally). Nature is what it is and the objective of scientific reasoning is to discover what Nature is.[4]

But this will not do at all for engineering. Engineering problems are overtly invented, and only rarely by engineers themselves. That is, engineering problems are explicitly given to engineers not by a supposed independently existing Nature, but by people who have, for a variety of generally obvious ulterior motives, invented them. Nor do the solutions to engineering problems exist in Nature, but only in the minds of those people who have posed, or assented to, the engineering problem as defined. To understand engineering, then, requires, in addition to understanding the cognitive methodologies employed by engineers in solving their problems, understanding the ineluctably willful and social processes by which engineering problems and solutions are specified, together with the circumstances of engineers' acquiescence in those specifications.[5] One thing is certain. Engineering, on the face of it, is very different form science as traditionally conceived, and as scientists continue to conceive it, because the intentionality informing the engineering problem specification process, and coordinately engineering reasoning, is very different from the presumptive intentionality informing the scientific problem specification process, and scientific reasoning.

As a matter of social fact, scientists and engineers constitute two distinct professional communities.[6] The contrast that I wish to draw here points to their distinctiveness as epistemological and metaphysical communities, as well. For scientists, intentionalities, value judgments, and willfulness may be acknowledged as influencing their professional conduct, but not as affecting the content of science. Thus, if one adopts a sociological interpretation of scientific knowledge, one must suppose that these factors operate implicitly. For engineers, on the other hand, intentionalities, value judgments, and willfulness explicitly shape the content no less than the conduct of their work. The upshot in either case is that scientists and engineers constitute two communities of practitioners whose mutual distinctiveness may well lie, not in their respective relationships to a *common* body of knowledge, but in the respective criteria each explicitly employs in constructing its *own* body of knowledge.

The claim that engineering knowledge is *sui generis* is reinforced by the implausibility of the supposition that scientific knowledge and engineering knowledge have a common object. Setting aside for the moment recent interpretations of science by defenders of the so-called Strong Program in the sociology of science,[7] the object of scientific knowledge, at least for scientific practitioners themselves, remains the impersonal, object-like value-free world that it was for the founders of modern science. This notion is sufficiently believable that the general public and many professional intellectuals, among them many historians, philosophers, and sociologists of science assent to it. Furthermore, even for those (and I count myself among them) who argue that scientific theories are socially constructed, it must be acknowledged that the practice of science explicitly, often aggressively, denies the social construction claim. In one way or another, scientists behave as if the object of scientific reasoning were independent of the reasoning process of the reasoners and of the social determinants of their own behavior. This in no way prejudices the case for the socially constructed character of scientific knowledge, which in any event needs to get behind the manifestly objective "public face" of science. Nor would accepting the social construction interpretation of scientific knowledge automatically vacate the contrast between science and engineering simply because they both would be social processes: they would, after all, remain two different social processes, distinguished by the contrasting commitments of their practitioners.

The object of engineering, however, appearances to the contrary notwithstanding, is a very different *type* of object than this. On no account could the object of engineering knowledge, or the specification of engineering problems, or the determination of what will count as acceptable solutions to them, be claimed even by engineers to be impersonal, value-free, or a matter of fact. From beginning to end and every step in between, engineering theory and practice is shaped by manifestly arbitrary, that is, explicitly willfully imposed, constraints expressive of a host of personal, institutional, social (including economic and political), and cultural (including aesthetic and religious) value judgments. The role of design in engineering illustrates this.

Design may or may not be the essence of engineering, but it is unquestionably central to engineering, and design is an explicitly valuational activity, a necessarily non-unique synthesis of a "box" of means, a set of imposed constraints, some natural, most arbitrary, and a fuzzy vision of an end to be achieved. A number of students of engineering, among them Eugene Ferguson, Brooke Hindle, Edwin Layton, Otto Mayr, and Anthony Wallace have argued that engineering design is distinctive *vis-à-vis* scientific analysis.[8] For example, while the latter is logical and discursive, the former, while no whit less intellectual, is non-logical, or better, not logical in the same sense, and non-discursive. Ferguson, Hindle, Thomas Hughes, and Samuel Edgerton have called attention to the role of visualization in design, and Ferguson in particular has argued that an over-emphasis on scientific analysis has corrupted the undergraduate engineering curriculum and is responsible for the proliferation of bad designs.[9] Others have argued that the strong science orientation of graduate engineering education and academic engineering research has harmed the teaching of design because design is not analytic, and it has reduced the effectiveness of engineers in industry because it discourages interest in manufacturing problems.[10]

In his monograph, *The Shape of Time*,[11] George Kubler compares engineering design solutions – solutions to problems posed by the desire to make new or improved artifacts – to aesthetic solutions to artistic problems. In addition to a visual component in design, Kubler notes that a kind of aesthetic judgment is operative in design. This judgment is embodied in the engineer's having to synthesize, in a consciously non-unique way, traditional and parallel contemporary solutions to this sort of problem; alternative solution options offered by available knowledge; and the spectrum of imposed constraints that enter into the very definitions of engineering problems and of what will count as acceptable solutions to them. Some of these constraints shaping what engineers know and how their knowledge is to be used include performance, cost, schedules to be met, quality, durability and reliability expectations, materials and production processes to be used or avoided, personnel availability, managerial prejudices, and corporate styles, among many others. None of these constraints, of course, has intrinsic weighting measures.

The synthetic and non-verbal aspects of engineering design contrast sharply with the analytical and discursive character of science. Together with the necessarily non-unique character of any given design solution, they entail profound methodological distinctions between science and engineering. Billy Vaughn Koen has attempted to define the engineering method in terms of intuitively derived and intuitively applied heuristics that are "in the final analysis unjustified, incapable of [strictly logical] justification and fallible." Their use generates "possibly, but not necessarily correct ways to solve a problem."[12] Layton has called design an instance of "ill-defined problem solving in which no solution algorithm is possible, no unique final solution state exists nor any path to it," and no "correct" solution exists either.[13] Hindle writes that the essence of design is elusive, "it is never fully subject to logic or rule" and "there is no deterministic or scientific way to design a machine."[14] Worse still, from the perspective of the epistemology of the physical sciences, it is a commonplace of engineering practice that what engineers working on a problem identity as the "best" solution to it is frequently rejected in favor of a less elegant solution on extra-technical grounds, for example, cost, marketability, style, legal restrictions.[15]

The non-uniqueness of engineering solutions has a number of consequences that distinguish engineering from science. Unless they are inter-translatable, rival scientific theories do not coexist well. The posture of the scientific community is that at any given time there is one best theory for explaining a particular phenomenon. The "solution space" of engineering problems is far richer and more complex than this. As a result, to attempt to explain why engineering solutions replace one another, why artifacts, machines, and technologies evolve as they do, will be very different from attempting to explain why scientific theories replace one another as they do, even if scientific theories *are* socially constructed. That is, even on the latter view, different sorts of social influences will be identified as operative.[16]

Again, the consciousness on the part of engineers that the specification of their problems and the terms of acceptable solutions to them are determined by arbitrary constraints imposed on them from outside engineering constitutes a fundamental distinction of engineering from science. This distinction holds even if the radical sociologists of science are correct and the content of science, no less than its conduct, is shapes by social and historical factors, so that the classical notion of objectivity is a myth. Practicing scientists do not experience the object of their research as defined by the wills of other scientists, let alone by non-scientists, yet this is the norm for engineers. That is, it is the norm for engineers, the overwhelming majority of whom are employees in private industry, that the constraints they must satisfy come from outside engineering: from managerial interpretations of the marketplace, of institutional needs, of political objectives, or of corporate agendas to which engineering excellence as defined by engineers is typically a secondary consideration.[17]

The preceding discussion at least suggests that the characterization of engineering as applied science is misleading because engineering knowledge and its object are distinctive by comparison with scientific knowledge and its object, and because engineering reasoning is distinctive, too. If scientific knowledge and engineering knowledge embody distinctive rationalities, and if the world that is the object of scientific knowledge is different from the world that is the object of engineering knowledge, then engineering deserves our attention and for two reasons. Engineering poses intellectual problems parallel to those posed by science, problems that have been ignored because of the pervasive underestimation of engineering in Western "high" culture. And engineering, as an element of technology as a social process, is actually changing the world.

The distinctiveness of engineering knowledge *vis-à-vis* traditional epistemology and philosophy of science is that it is a form of knowing in which knowledge claims, ostensibly validated by their "successful" implementation in the form of technologies (artifacts, production processes, or techniques), cannot be separated from the willfulness embedded in selective interest, selective action, and the exercise of power.[18] Engineering knowledge is a form of knowledge that develops in the context of an explicitly constrained decision-making process characterized by such parameters as Herbert Simon's notions of satisficing and bounded rationality,[19] by parochial value judgments reflecting the agendas of the institutions within which engineering is practiced and finding expression in the formulation of engineering specifications, in personnel utilization decisions, in issues of control, the choices of ends and of means acceptable for achieving them, in determinations of authority and accountability.[20] Engineering knowledge is distinctive for being driven by the search for consciously non-unique solutions to explicitly invented problems.

Unlike science as traditionally conceived and still commonly characterized, especially by scientists themselves, neither engineering knowledge nor engineering practice can be separated from the needs, values, and institutions of the communities within which engineering is pursued. These enter into engineering as endogenous factors, while the science community continues to maintain that for science these are only exogenous factors, that basic research is in principle an objective pursuit of the way things are "out there," in the objectively existing world.

Engineering thus poses a new set of epistemological problems deriving from a rationality that is different from that of science because the rationality of engineering involves volition, is necessarily uncertain, transient and non-unique, and is explicitly valuational and arbitrary, Furthermore, engineering also poses a distinctive set of metaphysical problems. The judgment that engineering solutions "work" is a social judgment; sociological factors must be brought directly into engineering epistemology and ontology.[21] The fact that engineering practice entails action on a world also implies that the ontology of engineering cannot be bracketed, so to speak, as the ontology of science can be, for example, under positivist interpretations. Engineering does not lend itself to a suspension of metaphysics in the manner of a coherence theory of truth: realism of some sort would seem to be intrinsic to engineering.

Jerome Ravetz and Bruno Latour have commented on the co-evolution of scientific facts, ideas, methodologies, and tools; also, that scientific instruments in some sense create the phenomena they were designed to measure.[22] Andrew Pickering offers a detailed account of how neutral currents in elementary particle physics emerged, as an "observed" phenomenon, out of a creative interaction between theoretical anticipation and available data.[23] But in engineering this interdependence among the object of thinking, the subject of thinking, and the concrete expressions of that thinking is so manifest that it will not permit the kind of idealization of engineering reasoning that has been imposed on scientific reasoning. The technologies produced with the aid of a body of technical engineering knowledge actualize realities that seem to be of a different order than those projected by science.[24] Crudely put, science has as its central activity generating descriptions of the way things are. Engineering has as its central activity generating things that have never been.

What the metaphysical commitments and implications of engineering are need to be investigated, but the recognition of the fundamental willfulness of technology was of sufficient moment for Martin Heidegger that he inverted the received tradition and argued that science was both logically and ontologically subordinate to technology.[25] For Heidegger, engineering is expressive of a will to make the world be for us as we want it to be. The will to engineer, so to speak, creates social and intellectual "spaces," respectively, for technology and for science. As a consequence, science's distinctive conceptual commitments ultimately, and unconsciously, serve engineering's needs, rather than the other way around.

It is possible to build further on these observations and to begin to identify the distinctive world outlook of engineering, what I earlier called its *theoria*. This would be to the *theoria* of the physical sciences as induction is to deduction, as the pragmatic is to the rational, as non-deterministic design is to deterministic law, as the contextual is to the universal, as the pluralistic and contingent is to the unique and necessary, as the open-ended is to the once-and-for-all, as Ciceronian rhetoric is to Platonic philosophy, as the skeptical is to the certain, as the Principle of Insufficient Reason is to the Principle of Sufficient Reason.

The contrasts set up here overlap and are largely self-explanatory. The world view of modern science is captured by a cluster of descriptors that includes deductive, Rational (in the classic sense of the term), deterministic, lawful (or law postulating), universal, necessary, uniquely true, timeless. Again, these terms describe science as its practice reveals that its practitioners conceive it, which is not necessarily science as it "really is." Quantum mechanical indeterminacy does not controvert the generic determinism, necessity, or certainty of science, for the equations of quantum mechanics still evolve deterministically, even though certain of the terms of those equations may be interpreted as probability distributions. The current prominence of chaos theory – as much a misnomer as relativity theory, in that the latter no more justifies the inference that everything is relative than the former does that natural phenomena are anarchic – reinforces this, as its practitioners busily formulate the laws governing the transition from orderly to only apparently disorderly behavior. Science is Platonic (or perhaps more accurately Pythagorean) in its intellectualism, in its embrace of the Principle of Sufficient Reason, and in its practitioners' pursuit of idealized, mathematical abstractions as the "true" reality from which the physical world can be deduced.

Engineering, by contrast, embraces what I have elsewhere called the Principle of Insufficient Reason that, as a matter of fact, dominates our daily lives.[26] That is, we continually experience the need to make deliberate decisions that cannot be made the conclusions of deductive arguments. However thoughtfully we try to act, we never possess information sufficiently complete and sufficiently general to permit acting on logically justifiable grounds, where "logically" means deductively. Soren Kierkegaard once wrote: "It is perfectly true as philosophers say, that life can only be understood backwards. But they forget that other proposition, that it must be lived forward."[27] Similarly, Moses Maimonides noted that retrospectively, all natural phenomena look necessary, while prospectively they appear contingent.[28] *Mutatis mutandis,* science (as practiced) concerns itself with the construction of best necessary explanations, while engineering generates a range of inevitably contingent options for action in real time, Kierkegaard's "living forward," which entails the superaddition of will to reason. The prominence of the Principle of Sufficient Reason in the Western intellectual tradition, and the fact that the Principle of Insufficient Reason has been taken seriously only by the minority community of skeptics, probabilists, and pragmatists, are further evidence of a bias toward science in Western culture.

The non-deterministic logic of design, and the non-uniqueness of the specification of engineering problems and solutions, entail that engineers must be guided by just such a Principle of Insufficient Reason. This makes engineering akin to the rhetorical tradition of the Sophists and of Isocrates, and especially of rhetoric as codified by Cicero and revived in the Late Middle Ages, rather than to classical philosophy, whether Platonic or Aristotelian.[29] The Humanist revival of rhetoric saw in it a methodology for creating a "public space" for disputation in which controversial issues could be argued non-coercively, a space in which convincing but not logically compelling inferences were defended, in the expectation that no one conclusion would be uniquely acceptable to all and that with accumulating knowledge and experience, every conclusion was vulnerable to revision.[30] Engineering shares many of these same characteristics and a philosophical inquiry into engineering would have to overcome the condescension with which philosophers in the modern period have treated rhetoric.

Finally, while engineering poses questions of truth and reality analogous to those posed by science, it also draws into their consideration questions of justice, of the good, and of beauty: if not in the artistic, then in the philosophic sense of judging that the balance of means and ends is harmonious. This is entailed by the acknowledgment that engineering is a *praxis*, which, as Paul Goodman argued twenty years ago, makes engineering a branch of moral philosophy.[31] Because the practice of engineering always takes place in action contexts, and because the form of its characteristic rationality includes a range of synthetic, and personal, and social-volitional elements, the practice of engineering, unlike the practices of science, mathematics, and philosophy, is inseparable from questions of *right* action. And in distinct contrast to science, the problems engineering poses for philosophy have a holistic character. The epistemological and methodological questions raised by engineering knowledge cannot be separated from questions of engineering's ontology because engineering entails action on a *world*. And because engineering entails *action* on a world, engineering intrinsically and inescapably raises moral, political, and aesthetic questions that traditionally are external to the conduct of science.

But these intellectual justifications for studying engineering carry in their train powerful obstacles to such study. The intellectual underestimation of engineering is merely a symptom of deep Western cultural prejudices that are likely to be difficult to overcome. And the fact that engineering is implicated in changing the world means that it is implicated as well in the struggles of powerful vested interests acting to control, and to benefit from, that change. There is little incentive for these interests to encourage public scrutiny of the technological innovation process and of the role played in it by engineering, and every reason to suppose that they would wish to discourage it.

I have referred on a number of occasions to a Western cultural prejudice against engineering. What evidence is there for such a prejudice?

One might begin by asking why even the educated public knows so much less about engineering and about engineers than they do about science and scientists. The National Academy of Engineering notwithstanding, very few of the people who could name five or even ten great scientists could name two great engineers, and one can confidently anticipate that the leading candidates would be Thomas Alva Edison and Henry Ford.

The recent crusades on behalf of public literacy about science and technology focus almost exclusively on science and the science component within technology, for example, scientific breakthroughs in materials research that underlie potential superconductor-based technologies. Popularizaticns in the vast majority of cases are of scientific theories – cosmology and astrophysics, relativity and quantum theories, plate tectonics, recombinant DNA – or of the careers of scientists, or of the wonders promised by new technologies and the scientific research that went into their discovery and will have to be done before they can be perfected. Has even one episode of NOVA attempted to explain to the public the role of engineering in technology, or the fact that engineers, too, engage in knowledge-generating research?

There is, of course, no Nobel prize for engineering achievements and no National Engineering Foundation to publicly channel public funds into engineering research. It is therefore hardly surprising that technological accomplishments are almost invariably ascribed to science primarily and engineering secondarily. This is a cliché as regards such disparate enterprises as the space program and industrial production. Space successes are scientific triumphs and our industrial competitive strength is our basic science research apparatus. Space disasters are engineering failures, as is our inability to compete with the Japanese. It is worth recalling, too, that the World War II successes of the Office of Scientific Research and Development, while ascribed to science and used as a rationale after the war for permanent mechanisms, including the National Science Foundation, for public funding of scientific research, were largely engineering triumphs, as Vannevar Bush himself somewhat ruefully noted.[32] It is not irrelevant to ask if the very choice of the name "Office of Scientific Research and Development" did not reflect a conscious decision to capitalize on a higher public regard for science than for engineering, a question that could with some justification be asked of the National Science Foundation and the National Academy of Sciences as well.[33]

A good elementary school staff today invariably includes science specialists with curricular materials adapted to the abilities of children from kindergarten up. The high school curriculum invariably includes the opportunity to take at least three years of science, and in a growing number of states three years of science is a requirement for graduation. Virtually all undergraduate liberal arts degree programs require that students take some courses in mathematics and/or the physical or life sciences. At no point in this sixteen-year span of education is there perceived to be a need to introduce students to engineering, even in those curricula that acknowledge technology as a formative social influence. That is, even where science, technology, and society (STS) materials are included in high school and college curricula, the engineering dimension of technology is commonly treated as a "black box." The Council for the Understanding of Technology in Human Affairs, and the Sloan Foundation program for the New Liberal Arts, have had very limited success, at best, in bringing engineering (as distinct from the social relations of science and technology) into the coursework of students other than engineering majors.

Cecelia Tichi has called the engineer "the invisible man of [academic] American studies."[34] In spite of the enormous social implications of technology for American society in the period from the Civil War to the Great Depression of the 1930s, and in spite of the popular identification at that time of the engineer as the agent of technological innovation, American studies has largely ignored engineers and engineering. So has American popular culture, except for a period of some fifteen years from the late 1890s on, during which engineers regularly appeared as the heroes of best-selling novels and silent films.[35] Indeed, until the 1960s, academic intellectuals also ignored science and technology, or subordinated them to political, economic, and social-psychological agents of social change.

When, in the course of the 1960s, Western intellectuals initiated a broadly based, systematic critique of the concept of objectivity, the result was a renovation of the history, philosophy, and sociology of science.[36] A consensus grew, and has since been sustained, that rather than being value-free, culture-free, and ahistorical, that is, "objective," scientific knowledge has a fundamentally social-historical character. This has proven to be a powerful stimulus to the study by humanists and social scientists of the nature of scientific reasoning, the bases of scientific theory replacement, and the synchronic and diachronic social relations of science. The fulcrum of the new perspective on science has been a commitment to the value-laden character of all theory and thus of all knowledge, including scientific knowledge. Coordinately, the objectivity of technology was also criticized, and rejected. As a result, the study of the social relations of technology – including its history, especially its social history, its sociology, politics, public policy, and social impact – has enjoyed a continuing popularity, though philosophy of technology has not fared nearly as well as these other disciplines.

What did not emerge from these intellectual developments was the philosophy of engineering as an accepted area of specialization within philosophy, or of sociology of engineering; there has been some limited interest in the history of engineering as a specialty within the history of technology, especially the history of engineering professionalism and education. Today, then, philosophy of science is a fully accepted and highly respected branch of philosophy, while philosophy of engineering carries as much professional distinction as philosophy of parapsychology. Yet the critique of objectivity that sparked the renovation of science and technology studies, pivoting as it does on the value-laden character of science and technology, applies even more forcefully, and more evidently, to engineering than to science, as the preceding discussion of design indicates. Indeed, although I cannot argue it here, the revaluation of objectivity seems to me to imply that philosophy of engineering should be the paradigm for philosophy of science, rather than the reverse.

Finally, engineering is caught up in high-culture prejudices against the particular, against concreteness, practice and action, and in favor of the universal, abstraction, theory and contemplation that go back to Pythagoras, Plato, and Aristotle.[37] These prejudices triumphed in the struggle between the Sophists and the Socratics for the definition of philosophy, a triumph reflected in the routine subordination of Roman culture by high-culture critics. They are strongly characteristic of medieval Scholasticism and Renaissance "magical" nature philosophy, and they persisted in modern science in the superior status accorded to theory over experiment.[38] The early assimilation of science and mathematics into the university curriculum contrasts sharply with the nineteenth-century battles over university-level status for the study of engineering, while engineering education and professionalization in the last hundred years have developed by deliberate emulation, or assimilation, of scientific research norms, institutions, and values.[39]

This context of cultural prejudice reinforces the rationale for the hierarchical ordering of science and engineering, namely, that science is more fundamental than engineering. But that science is superior to engineering because engineering can only be approached through science is, for one thing, historically insupportable.[40] Even during the Golden Age of engineering – the nineteenth and early twentieth centuries, when Western societies truly *were* revolutionized by new production, transportation, communication, construction, and energy technologies – engineering's indebtedness to science was modest at best, and in earlier transformations rung on European societies by technology, from the eleventh century to the nineteenth, the indebtedness was nonexistent.

Part of the problem of disentangling science, engineering, and technology lies in an equivocal use of the term "science." Engineering always embodies knowledge and if "science" is a synonym for "knowledge," then of course science is a precondition of engineering. But we have already seen that the kind of knowledge engineers wield is different from the kind that scientists wield, although some concrete expressions of their respective types of knowledge – functional relationships, data, instrumentation – are often transferrable. The crucial point to appreciate is that engineering on its own activity generates knowledge. It does not passively wait for knowledge to be given to it from a different community of practitioners in order for it to attempt increasingly complex enterprises.

This was the case with the development of wind – land and water – power technologies from the eleventh century on, and the development of increasingly sophisticated gear-driven machinery especially after the fifteenth century. The development of the factory system and early industrial chemistry, canal building, iron and steel metallurgy, bridge and tunnel construction, highways, the whole range of steam-power technologies among many, many others illustrate the point: engineering became increasingly sophisticated from the sixteenth century on, and especially from the late eighteenth century, on its own, not because it was being pushed by science.

This is not to say that engineering was then, or is now, independent of science. Modern science and modern engineering are collateral expressions of Western European cultural developments since the sixteenth century.[41] Furthermore, although then as now scientists and engineers constituted two distinct communities, they have not been isolated from one another. The possibility for mutual fertilization and systematic interaction existed and began to be exploited in the late eighteenth century. This interaction finds expression in a growing intensification of mathematics in engineering practice, and in an intensifying commitment to a systematic approach to engineering training, resting on experimentation and the appropriation of specific scientific results.

All of these were incorporated into engineering education in the course of the nineteenth century.[42] Concurrently, systematic interaction between science and engineering was stimulated by the emergence of the industrial research laboratory, but until the middle of the nineteenth century there were few scientific theories fertile enough to have practical applications. On the contrary, an autonomous drive within engineering in the context of the rush of contemporary technological innovations was pressing transportation, communication, power, and a wide range of production and materials processing technologies, independent of the rate of development of scientific theories bearing on those applications. James Clerk Maxwell's electromagnetic theory, after the experimental demonstration of electromagnetic waves by Heinrich Hertz and by Oliver Lodge, suggested Guglielmo Marconi's first spark gap transmitter and receiver. But the development of a practical radio communication technology proceeded largely in advance of, and can even be said to have stimulated the development of, theories of tuning, signal propagation, and electronic amplification.[43]

The actual relationship of engineering to science is further complicated by the experience of the past half-century. During this time, the practices of science and engineering have become mutually interdependent, the politics of science and engineering in relation to society have become highly ideological, and the professional communities of scientists and engineers have developed in distinctive ways. In addition, we have become accustomed to associating engineering with doing, so that, on the one hand, we incorrectly conflate engineering and technology, and on the other, we speak, imprecisely, of scientists practicing engineering when they construct the most complex of their instruments.[44]

The complexity of their relationship notwithstanding, the central questions bearing on the claim that engineering is applied science are these: is science logically prior to engineering; is science chronologically prior to engineering? Both the historical record and the preceding arguments regarding the distinctiveness of engineering knowledge and the world that is its object suggest that the answer to both of these questions is No.

Engineering is prior to science, understood as the systematic, abstract study of natural phenomena, and this remained the case even in the era of the steam-power-based Industrial Revolution and into the eras of electricity, synthetic materials, radio, and flight. Had science stopped dead with Pierre Laplace, the societal transformations that took place in the nineteenth century would not have been affected. Part of the reason for this is that engineering has its own knowledge base. Under the influence of the prevailing prejudice, we are inclined to suppose that the knowledge components in medieval and Renaissance engineering, for example, were proto-scientific, or scientific "inclusions" within engineering, later reinforced by explicit borrowings, a process systematized by the progressive scientization of engineering education in the course of the nineteenth century. But if engineering knowledge is *sui generis*, then engineers are not fundamentally indebted to scientists for the knowledge they need to practice engineering.

In the absence of an institutionalized and fertile science, engineers generate their own theoretical knowledge in the course of solving their problems. They have been doing so for millennia, and continue to do so today. Furthermore, as Layton has shown, even where engineers did explicitly borrow from science, in thermodynamics and in electrical theory, for example, they had to rethink the abstract scientific knowledge and reformulate it as concrete engineering knowledge.[45] If we then add to this the distinction between the object worlds of scientific and engineering knowledge, the conclusion must be that science is not either chronologically or logically prior to engineering.

It is, nevertheless, easy to overstate the distinctiveness of science and engineering. While they *are* distinct in principle and at a certain level of practice, at other levels their practices overlap to such a degree that it seems pointless to attempt to distinguish them, while their professional institutions have been intertwined for more than a century. Engineering education especially has been based on a scientific ideal for engineering practice since the late nineteenth century, when a last effort to keep engineering training rooted in shop and job training failed to win wide support.

Engineering practice and the professional institutions of the engineering community have also been "scientized." Especially since the Second World War, engineering research, modelled after scientific research, has become a dominant force in academic engineering. The Ph.D. in engineering has become increasingly significant in private industrial research and development, and in engineering management. Extremely uncommon prior to World War II, and awarded in the early and mid-1940s at the rate of approximately a hundred a year, the number has grown to an average of three thousand a year over the last fifteen years. This is a far larger number than can be absorbed by educational institutions, yet industry has never defined a specific need for engineers with doctorates, other than organizational stature in R&D or R&D-management settings. The assimilation of the science research model to engineering has been so complete, however, that NSF was forced, in the early 1980s, to acknowledge engineering research as a funding obligation parallel to scientific research. This commitment was formalized by a change in the NSF charter in 1985 and by expanded NSF funding for engineering and the creation of a national network of Engineering Research Centers. Some, however, see the scientific model as having deflected engineering away from an orientation to manufacturing, with negative consequences for U.S. industrial competitiveness.[46]

At the same time that engineering institutions and practice have been reconstructed in emulation of scientific institutions and practice, science, as Jerome Ravetz has argued, has been industrialized. In part this has been the result of science's direct interaction with engineering in industrial settings, in part through its co-option by industry and government as a driver of economic growth and political and economic power, in part through the steadily increasing dependence of "big" science on highly sophisticated, very large and/or very expensive machinery and instrumentation.[47] That it is difficult to distinguish applied science from engineering in practice has potentially profound implications for "pure" science as well as for engineering, particularly as the distinction between pure and applied science is relative rather than absolute. Indeed, as suggested above, social construction interpretations of science imply that the covert *logos* of scientific reasoning is a variant of the *logos* of engineering reasoning, so that the real distinctiveness of science *vis-à-vis* engineering lies in the ideology motivating its overt objective facade.

There are, clearly, strong structural similarities today between the practices of science and engineering and these have been variously identified by Ravetz, Trevor Pinch and Wiebe Bijker, David Dickson, and most recently by Latour.[48] Layton has argued that the notion of the relationship of engineering to science prevalent at a given time expresses ideological commitments,[49] and this certainly seems to have been the case in the post-World War II period when the creation of the National Science Foundation was being debated.[50] As became clear in the late 1970s, when the creation of a National Engineering Foundation was proposed, the hold of the pure/basic science community on the public's monies is fragile, so that there is a great deal at stake for science in reinforcing the cultural, epistemological, and professional primacy of science in relation to engineering.[51]

Williams Jennings Bryan once argued that all political questions are ultimately moral questions and, given that the moral intuitions of politicians are *a priori* no better than the moral intuitions of the general citizenry, all political questions must be open to the public.[52] In the same vein, one could argue that all technological questions, including what engineers ought to be doing and how they ought to be doing it, are ultimately political questions, because technological action is the end-product of a series of value judgments in which society is implicated from beginning to end. Lack of technical expertise is not a valid ground for excluding the public from participation in technology policy decisions. This should be self-evident. Members of Congress, presidents and their cabinet secretaries, the military high command, and corporate upper management in the overwhelming majority of companies, have no training in science or engineering, yet they set science and technology policies as a matter of routine. Excluding the public is strictly a matter of maintaining power relationships. Were the will there to do so, the same technocratic infrastructure that services political, military, and corporate decision-makers could provide their putatively value-free technical expertise to the public as well.[53]

But this is, of course, highly unlikely to happen. The expertise represented by General Motors' engineers, for example, is proprietary to GM not only because GM has chosen to pay for their knowledge, but because what they know, and how what they know is being selectively exploited at GM, are not in the least value-free. The rubric "proprietary information" serves as a buffer for the latter fact as much as for expert information protected. It is not so much that Ford and Chrysler, Toyota and Honda would benefit from this information, as that GM could be harmed by a general awareness of the innovation choices GM was making and why, or of what GM's engineers thought about managerial decisions relating to automation, fuel efficiency, safety, and product quality.

Engineering, through technology, is directly implicated in changing the world. It is, therefore, directly implicated in the struggles for power among the institutions and individuals in society who see threats to their interests in new technologies, or who see in it opportunities for expanding those interests. One consequence of this fact is that there is a vested interest among powerful forces in society in not having the public appreciate the role played by engineering in technological innovation. Consider, for example, Karl Marx's oft-cited Eleventh Feuerbachian Thesis: "Until now philosophers have only attempted to interpret the world; the point, however, is to change it."[54] This is certainly wrong.

While it may be true that prior to Marx philosophers were satisfied to interpret the world, the point surely is not to *change* the world, but to *improve* it. The history of technology is evidence that it is not all that difficult to change the world in many superficial and profound ways, generally unpredictably; but it is plainly very difficult indeed to improve it. The asymmetry between "change" and "improve" – every improvement entails a change but not every change is an improvement – calls attention to the necessity of antecedent value judgments in order to determine whether any given change *is* an improvement or not. That is, as so many critics of technology have noted, only a vision of ends to be achieved allows us to determine whether appropriate means to reach those ends have been selected and are being employed effectively.[55] Technologies, of course, are always and necessarily means, precipitating changes that are commonly made possible only by access to engineering expertise. The process by which new technologies are generated and then disseminated in society would become vastly more difficult to manipulate by technological actors if the public recognized the fundamentally political character of the innovation process and of the role played in it by engineering.

Technological innovation, like revolutionary politics, has as a central deliberate objective changing some things about the world, to the advantage of the innovators, while leaving others unchanged, for the same reasons. But whether technology-induced changes are improvements or not entails judgments that refer back to the value structure of the decisions underlying technological action. No more than the revolutionary in relation to political action can the technical in relation to technological action be itself the source of judgments of the value of the changes induced by that action. Furthermore, what new technologies leave unchanged, for example, with respect to the distribution of political power, social power, and wealth is revealing of the socio-political forethought that enters into technological action.

The purported value neutrality of the technical is thus an ideologically motivated stratagem. It serves to insulate from criticism the dynamical factors determining technological action at the expense of misrepresenting engineering. As Lewis Mumford has insightfully observed, the claim that there is a value-free world of objects over against which we humans stand, uniquely, as subjects, is itself a complex value judgment.[56] That it won widespread support in the seventeenth century is a fact about Western culture of enormous significance for the destinies of modern science and engineering, and for the forms of their exploitation. What it signifies is that the impact of new technologies reflects antecedent value commitments in order for that impact to be realized. The triumph of the machine in the nineteenth century was for Mumford a consequence of the social, political, and cultural mechanization of *people* in the course of the preceding two hundred years in the West, rather than the reverse. That is, the triumph of the machine was not the result of the coerced transformation of a society that valued every human as an end, not a means, merely because of the invention of machines with certain technical characteristics.

Mumford's critique of technology, which strongly echoes themes in Giambattista Vico's *New Science*, suggests that a philosophy of engineering would open out to philosophical anthropology. The creation of technical knowledge, the selective exploitation of technical knowledge, and the social consequences of that exploitation are mutually interrelated. Furthermore, the social context of technological action and the constellation of political, economic, legal, and cultural institutions and values extant at a given moment in a society's existence, determine the managerial agendas that shape the applications of technical knowledge. Engineering is therefore inseparable not merely from the concrete institutional contexts of its practice by individual engineers operating under particular managerial directives. It is inseparable as well from the broader social environment of which those contexts are expressions.

The philosophy of science was for a long time pursued as if the *logos* of science were separable from the history of science and from the daily practice of science; as if the essence of science were an abstract mode of reasoning and thus accessible to pure thought. The philosophy of engineering, however, cannot be pursued in this way. From the start engineering must be understood as integral to its social context, on which the products of its work act, often with unanticipated consequences, and to which the content and the conduct of engineering is a response. The philosophy of engineering cannot, therefore, be a technical specialty within philosophy, as philosophy of science has become. Philosophy of engineering forces an engagement with an ontology not of *the* world, but of *a* world that some people have chosen to make and which then makes all people over as people. It forces as well an engagement with an epistemology of explicitly willful reasonableness, and with the interpenetration of the technical with the moral, the political, and the aesthetic. Paradoxically, then, a critical approach to engineering forces a systematic philosophical response, acknowledging the integral unity of the problematics of the real, the true, the good, the just, and the beautiful within the domain of the technical.

Lehigh University

NOTES

[1] Steven L. Goldman, "The Social Captivity of Engineering," in P. Durbin, ed., *Critical Perspectives on Non-Academic Science and Engineering* (Bethlehem, Pa.: Lehigh University Press, forthcoming).

[2] See, for example, Ian Hacking, *Representing and Intervening* (Cambridge: Cambridge University Press, 1983); Hannah Arendt, *The Human Condition* (Chicago: University of Chicago Press, 1958), esp. pp. 289ff on the reversal of the order of priority of the *vita activa* and the *vita contemplativa* in the modern age; George Ovitt, *The Restoration of Perfection: Labor and Technology in Medieval Culture* (New Brunswick, N.J.: Rutgers University Press, 1987), chapter 4: "The Mechanical Arts in the Order of Knowledge," pp. 107–136; Steven Shapin and Simon Schaffer, *Leviathan and the Air Pump* (Princeton, N.J.: Princeton University Press, 1985).

[3] In Richard Popkin, *The History of Skepticism from Erasmus to Descartes* (Berkeley: University of California Press, 1979), p. 5.

[4] It is difficult to document the obvious, but only because no compact group of citations can do justice to its being obvious. On one level, contemporary rhetoric concerning black holes, gravitational lenses, plate tectonics, and unconscious mental processes, supports the realism of virtually all working scientists. More specifically, the continuing relevance of the traditional conception of science as objective, necessary, and universal is reflected in Peter Medawar's *The Limits of Science* (Oxford: Oxford University Press, 1984), and especially in the reviews of the book in scientific journals; in "Constraints on Construction" (*Social Studies of Science*, **16** [1986]:372–383), a rejection by physicists Yves Gingrich and Sylvan S. Schweber of Andrew Pickering's interpretation of quantum chromodynamics as a socially constructed theory (see note 7 below); and in P. Davies and J. Brown, eds., *The Ghost in the Atom* (Cambridge: Cambridge University Press, 1986), a collection of transcripts of interviews on the meaning of quantum physics with eight eminent physicists, every one of whom explicitly affirms his realism and objectivism. Coordinately, such sociologist theoreticians of knowledge as Karl Mannheim and Robert Merton, spanning the first two-thirds of the twentieth century, explicitly shelter the objectivity of the content of science from social determinants, while Sal Restivo continues to defend the objectivity of "true" science in principle, notwithstanding the socially determined character of modern science.

[5] See the excellent discussion of the diesel engine's transformation from idea to commercial product in Bruno Latour, *Science in Action* (Cambridge, Mass.: Harvard University Press, 1987), pp. 103–144. Also the discussion of the intentionality of engineering in Steven L. Goldman, "The Techne of Philosophy and the Philosophy of Technology," in P. Durbin, ed., *Research in Philosophy and Technology*, volume 7 (Greenwich, Conn.: JAI Press, 1984), pp. 130–134.

[6] See, for example, Robert Perucci and Joel E. Gerstl, *Profession Without Community* (New York: Random House, 1969), and Edwin T. Layton, "Engineering and Science as Distinct Activities," in J. Burnet, ed., *Technology and Science: Important Distinctions for Liberal Arts Colleges* (Davidson, North Carolina: Davidson College, 1984).

[7] For an overview of the Strong Program, see the introductory essay to Karin Knorr-

Cetina and Michael Mulkay, eds., *Science Observed* (London: Sage, 1983), pp. 1–18. For illustrations, see Harry Collins, *Changing Order: Replication and Induction in Scientific Practice* (London: Sage, 1985); Bruno Latour, *Science in Action* (Chicago: University of Chicago Press: 1984); and Thomas Pickering, *Constructing Quarks* (Chicago: University of Chicago Press, 1986).

[8] Samuel Y. Edgerton, Jr., *The Renaissance Rediscovery of Linear Perspective* (New York: Basic Books, 1975); Eugene Ferguson, "The Mind's Eye: Non-Verbal Thought in Technology and Science," *Science* **197** (26 August, 1977): 827–836; Thomas P. Hughes, "Model Builders and Instrument Makers," *Science in Context* **2** (Spring 1988):59–75; Brooke Hindle, *Emulation and Invention* (New York: Norton, 1983); Otto Mayr, ed., *Philosophers and Machines* (New York: Science History Publications, 1976), introduction; Anthony Wallace, *Rockdale* (New York: Knopf, 1978). Edwin Layton sums up the point perfectly: "From the point of view of modern science, design is nothing, but from the point of view of engineering, design is everything," in "American Ideologies of Science and Engineering," *Technology and Culture* **17** (1976); see also Layton, "Science and Engineering Design," in *Bridge to the Future* (New York: New York Academy of Sciences, 1984).

[9] According to Edgerton, Ferguson, and Hindle. See previous note.

[10] According to Ferguson, "The scientific approach to engineering is systematic, but intellectually impoverished," "The Mind's Eye," p. 833. That this was a source of concern earlier than the immediate past, see W. L. Everitt, "The Phoenix-Challenge to Engineering Education," in *Proceedings of the Society for the Promotion of Engineering Education* (1944):321–328; there Everitt argues that the engineering curriculum no longer teaches engineering because of its abstract and analytical character; also, J. H. Hollomon, *Future Directions for Engineering Education* (Washington, D.C.: American Society for Engineering Education, 1975), especially pp. 47–48, where Holloman cites the need for reintroducing "the art of engineering" into the engineering curriculum from which it has been squeezed out by basic science courses.

[11] George Kubler, *The Shape of Time: Remarks on the History of Things* (New Haven, Conn.:Yale University Press, 1962).

[12] Billy V. Koen, "Towards a Definition of the Engineering Method," in *The Bent* (of Tau Beta Pi) **76** (Spring 1985):28–33. This is amplified in Koen's *Definition of the Engineering Method* (Washington, D.C.: American Society for Engineering Education, 1985).

[13] Layton, "Science and Engineering Design" (note 8, above), p. 59.

[14] Hindle, *Emulation and Invention* (note 8, above), pp. 133 and 137.

[15] A primary objective of engineering education today must be "an understanding of social and economic forces and their relationship with engineering systems, including the idea that the best technical solution may not be feasible when viewed in its social, political, or legal context." See Edward T. Cranach, *Engineering Undergraduate Education* (Washington D.C.: National Academy of Engineering Press, 1986), p. 10.

[16] An excellent illustration of this point is contained in an article in *Science* **244** (June 30, 1989):1532–1534, comparing the aircraft control systems of the Airbus Industrie A320 and the McDonnell Douglas MD-II passenger planes. After reviewing the contrasting design philosophies embodied in these control systems, M. Mitchell Waldrop writes: "So, who's right? Maybe everybody. It's a cliché that engineering is

an art, but it is. And it's perfectly possible for Airbus, McDonnell Douglas, and all the rest to come up with very different solutions to the problem of aircraft automation, and still be perfectly correct."

17 Approximately 80% of U.S. engineers are employed in private industry, with large concentrations in a small number of corporations. For a discussion and references, see Goldman, "The Social Captivity of Engineering" (note 1, above).

18 Latour, *Science in Action* (note 5, above).

19 Herbert A. Simon, "The Logic of Heuristic Decision-Making," in his *Models of Discovery* (Dordrecht: Reidel, 1977), pp. 154–178; also, *The Sciences of the Artificial* (Cambridge, Mass.: MIT Press, 1969).

20 For a description of the impact of conflicting institutional agendas on the development and regulation of nuclear power in California, see Richard A. Meehan, *The Atom and the Fault* (Cambridge, Mass.: MIT Press, 1985).

21 Latour, *Science in Action* (note 5, above); see also Trevor J. Pinch and Wiebe E. Bijker, "The Social Construction of Facts and Artefacts: or How the Sociology of Science and the Sociology of Technology Might Benefit Each Other," *Social Studies of Science* **14** (1984):399–341.

22 Latour, *Science in Action*; Jerome R. Ravetz, *Scientific Knowledge and Its Social Problems* (Oxford: Oxford University Press, 1971), especially pp. 31–68 and 321–363.

23 Pickering, *Constructing Quarks* (note 7, above), pp. 180–195; but see also the review-critique of the book by Gingras and Schweber (note 4, above), for an objectivist response.

24 For example, Thomas Hughes writes that Nikola Tesla's "idealized images stimulated his creative endeavors. If the image did not coincide with the reality, he endeavored to reorganize the real so that it would approximate his ideal"; *Networks of Power* (Baltimore, Md.: Johns Hopkins University Press, 1986), p. 115.

25 Martin Heidegger, "The Question Concerning Technology," in *The Question Concerning Technology and Other Essays* (New York: Harper, 1977). On the ways that artifacts express the willfulness they embody, see Thomas H. von Laue, "Technology, Society and Freedom," *Technology in Society* **5** (1983).

26 Goldman, "The Techne of Philosophy and the Philosophy of Technology" (note 5, above), pp. 136–139; also, "On the Incommensurability of Science and Religion," *Working Papers Series*, volume 1 (Bethlehem, Pa.: Technology Studies Resource Center, Lehigh University, 1984), pp. 1–24.

27 Soren Kierkegaard in an entry for 1843 in his *Journals: Selected and Revised Edition* (London: Oxford University Press, 1938), p. 127.

28 Moses Maimonides, *Guide for the Perplexed*, 2, 17.

29 David Marsh, *The Quattrocento Dialogue: A Classical Tradition and Humanist Innovation* (Cambridge, Mass.: Harvard University Press, 1980), pp. 1–23.

30 *Ibid.*, pp. 24–37.

31 Paul Goodman, *The New Reformation: Notes of a Neolithic Conservative* (New York: Random House, 1970), pp. 3–23.

32 Nathan Reingold, "Vannevar Bush's New Deal for Research: Or, the Triumph of the Old Order," *Historical Studies in the Physical Sciences* **17**, part 2 (1987):299–344. See also George Wise, "Science and Tech," *Osiris*, second series (1985), volume 1, pp. 229–246.

[33] The National Academy of Engineering is itself some ninety years younger than the National Academy of Science.
[34] Cecelia Tichi, *Shifting Gears: Technology, Literature, Culture in Modernist America* (Chapel Hill: University of North Carolina Press, 1987), p. 98.
[35] *Ibid.*, pp. 99f; see also Steven L. Goldman, "Images of Technology in Popular Film," *Science, Technology, and Human Values* **14** (Summer 1989):275–301.
[36] The reconceptualization of objectivity in the 1960s seems to me a fundamental cultural development, one cognate with the mass anti-establishment movements of the period: civil rights, consumer activism, environmentalism, alternative life styles, opposition to nuclear power, to the Vietnam War, to multinational corporations, to the military and intelligence communities, and to science and technology. The shift from so-called internalist to externalist history of science and technology, the promotion of non-Western and non-rational modes of knowing, and the revival of relativism and subjectivism are symptomatic of this development, as was the rise of science, technology, and society (STS) studies.
[37] Steven L. Goldman, "On the Interpretation of Symbols and the Christian Origins of Modern Science," *Journal of Religion* **62** (1982):1–21; also, "From Love to Gravity: Renaissance Magical Nature Philosophy and the Origins of Modern Science," an unpublished paper available from the author.
[38] Lawrence P. Grayson, "A Brief History of Engineering Education in the United States," *Journal of Engineering Education* **67** (December 1977):246–264. Also, Steven L. Goldman, "The History of Engineering Education: Perennial Issues in the Supply and Training of Talent," a report prepared for the Office of Technology Assessment, NTIS #PB 88–177 951/AS.
[39] Goldman, "History of Engineering Education," especially the bibliography.
[40] The historical priority of technology *vis-à-vis* science has been central to Lewis Mumford's critique of technology, from *Technics and Civilization* (1934) to *The Myth of The Machine* (1967/1970).
[41] Among these are the intensification of secularism and materialism as expressed in the art, philosophy, political theory, and economic policies of the period.
[42] Grayson, "Brief History" (note 38, above), especially his many references.
[43] See, for example, Hugh G. J. Aitken, *Syntony and Spark: The Origins of Radio* (New York: Wiley, 1976), and *The Continuous Wave: Technology and American Radio 1900–1932* (Princeton, N.J.: Princeton University Press, 1985).
[44] On scientists as engineers, see Lillian Hoddeson, "The First Large-Scale Application of Superconductivity: The Fermilab Energy Doubler, 1972–1983," *Historical Studies in the Physical and Biological Sciences* **18**, part 1 (1987). Also, Joan Lisa Bromberg, "Engineering Knowledge in the Laser Field," *Technology and Culture* **27** (October 1986).
[45] Layton, "Science and Engineering Design" (note 8, above).
[46] See the discussion of U.S. industrial competitiveness in Part III of Goldman, "History of Engineering Education" (note 38, above).
[47] Ravetz, *Scientific Knowledge* (note 22, above), especially chapter 2, "Social Problems of Industrialized Science," pp. 31–68.
[48] *Ibid.*; also Pinch and Bijker (note 21, above) and Latour (note 5, above). See also David Dickson, *The New Politics of Science* (New York: Pantheon, 1984).
[49] Edwin T. Layton, Jr., "American Ideologies of Science and Engineering,"

Technology and Culture **17** (1976).

[50] J. Merton England, *A Patron for Pure Science: The National Science Foundation's Formative Years, 1945–1957* (Washington, D.C.: National Science Foundation, 1982), and Milton Rowan, "Politics and Pure Research: The Origins of the National Science Foundation, 1942–1954" (doctoral dissertation, Miami University, Oxford, Ohio, 1985).

[51] See, for example, Lewis M. Branscomb, "Engineering and the National Science Foundation," *Science* **224** (27 April, 1984).

[52] Cited in Richard Hofstadter, *The Age of Reform* (New York: Vintage, 1955), p. 262.

[53] Sheldon Krimsky seems to apply to the public a stricter standard of knowledgeability as a condition for participation in technology policy decisions than to policy makers themselves in his essay, "Epistemic Considerations on the Value of Folk-Wisdom in Science and Technology," *Policy Studies Review* **3** (February 1984):246–261.

[54] *Die Philosophen haben die Welt nur verschieden interpretiert, es kommt darauf an, sie zu veranderen*; Karl Marx, *Werke*, Band 3 (Berlin: Dietz Verlag, 1954), p. 7.

[55] For example, Robert S. Morison, "Visions," in A. Teich, ed., *Technology and Man's Future* (New York: St. Martin's Press, 1981), pp. 7–22.

[56] Lewis Mumford, *Technics and Civilization* (New York: Harcourt, 1934), pp. 361ff.

JOSÉ SANMARTÍN

ALTERNATIVES FOR EVALUATING THE EFFECTS OF GENETIC ENGINEERING ON HUMAN DEVELOPMENT*

On the subject of genetic engineering, as with most technological developments, speculation often outruns reality. What I mean to say is that genetic engineering has its share of fortune tellers, and their prognostications are not limited to justifying the current state of the technology; they go on to demonstrate how extensions to unheard – of dimensions will demand a new and better social order. The term "revolution" – as everyone knows who is familiar with the history of science, of technology, and of culture generally – has been applied since the nineteenth century not only to political upheavals but also to technical innovations. Furthermore, anyone thus enlightened also knows that such revolutionary promises, in the long run, often lead to quite modest results.

In fact, all major technological innovations have, historically, been cheered on by a chorus of voices praising their excellence. And the chorus is joined, oddly enough, not only by engineers and technicians but by journalists, philosophers, humanists, and scientists – particularly those in the human sciences and the basic sciences. The words they sing typically have little to do with the material advantages of particular technologies; rather, they highlight the contribution of each particular innovation to bringing about a new social order which will be more just, more democratic, or more egalitarian. In short, each technological revolution is depicted as ushering in a utopia newly put within human reach.

It is not uncommon today for the chorus cheering on technological development to include within the predicted bucolic utopia great claims for both personal computing and genetic engineering.

From the days of Francis Bacon's *New Atlantis* on – again, as everyone knows – utopian technological predictions have turned out to be pipe-dreams never realized. I have dealt elsewhere with the question of responsibility for this sort of fiasco.[1] Some say that critics of technological development, making precipitous negative judgments, are really the ones guilty of the social outrages. I ask instead whether the responsibility might not lie with human nature itself: are humans capable of taking full advantage of their

Paul T. Durbin (ed.), Broad and Narrow Interpretations of Philosophy of Technology, 153–166.

achievements, including technological achievements, or even of avoiding the worst negative consequences of those achievements?

It seems to me undeniable that humans are the product of their own technology.[2] But that does not mean that certain aspects of what it means to be human are not products of natural selection. And it may well be these aspects that make us incapable of taking full advantage of our achievements. Some critics even go so far as to say that these natural characteristics, out of our control, are what make us, in the end, a threat to our own survival.[3]

Today the most powerful politician, capable of launching nuclear missiles with the flick of a finger, is, from this technocritical perspective, not very different from the caveman with hypertrophic aggressions uncontrollable by natural devices – as certain doomsaying anthropologists have warned.[4] According to such critics, these are the traits that distinguish human nature. A wolf's aggressiveness is limited by natural factors – for instance, the bared neck of a vanquished adversary. Its aggressiveness is natural, triggered by mechanisms of natural selection and, in consequence, limited in the same way. Human aggressiveness, in contrast, is exercised with instruments, which is to say it is a matter of culture; and nature by definition has little or nothing to do with aggressiveness expressed by artifice, with tools, in a cultural fashion. Only cultural devices can counter or control cultural products.

If these pessimistic anthropologists are correct, we have a serious problem. We humans are, paradoxically, aggressive by nature. Aggression even plays a role in human evolution. We have learned to exercise our aggressiveness in cultural, not natural, ways. Our evolution has made it possible for human aggression – both it evocation and its control – to escape nature. The balance between aggression and the mechanisms for its control that operate in nature has been upset in the case of humans.

Human aggressiveness, using tools, has become hypertrophic. We kill by throwing stones, then spears; by shooting arrows, then bullets; by dropping bombs, then launching missiles – each step involving ever greater distances. This separates active from passive agency; it eliminates any possibility of natural limits. As aggressiveness becomes hypertrophic, it becomes less animal and more human – more "cultural," if you will.

So at last we come to the point: genetic intervention. We are dangerous to our own well being. We are, at bottom, *Homo australopithecus* (or something similar), left at the mercy of our own aggressiveness, with a missile on our shoulder and nothing else to prevent the inevitable. We may have a sense of "responsible rationality" to rely upon, but recall that it has failed again and again throughout history to prevent large-scale war and similar disasters. We can then take the final step in the argument. If we assume a biological basis for aggressiveness and its inevitable hypertrophy through cultural mechanisms, and, furthermore, if we assume that all previous attempts to eradicate aggression have been unsuccessful, is it not time to conclude that, for the sake of the survival of the human race, we need to intervene biologically? Do we not need to eliminate the threat at its root? A dead dog can no longer bite, so to speak.

Next comes a substitution: this possibility is transformed into necessity – the necessity of carrying out the possibility. Our situation, some say, is so worrisome today, our aggressiveness so hypertrophied, and the situation so close to disaster, that it seems difficult to bank on conscience, on mere scruples, to prevail over what is so foreboding. It is clear that those who think along these lines are convinced that aggressiveness is biologically determined in the strict sense. Given this assumption, plus the tools needed to intervene in the biological mechanisms of this determinism, they arrive at their conclusion: it is *necessary* to choose between two extremes, slaughter and salvation. Biology leads to catastrophe. Salvation can be attained by cultural means – by a culture that has enabled us to develop instruments to intervene in our own biology, to cure what is pathological there. What lies between slaughter and salvation is merely scruples – so they add.

Furthermore, what has been said about aggressiveness can be extended – by such technophile futurists – to other important aspects of human social behavior. It seems possible, if only we will use the appropriate instruments, to eradicate such undesirable features of human behavior – detrimental as they are to the survival of the human race – as xenophobia, sexism, chauvinism generally, even egocentrism. The idea, then, is to modify human nature through the use of technical devices, acting directly on the biological basis of those hypertrophied aspects of human behavior that have become, by the very reason of their hypertrophy, a threat to human survival.

And I emphasize once again that, for these people, the question boils down to a choice between that technology which can supply us with something as precious as the improvement of human nature and mere scruples. Here is Jonathan Glover on the matter:

Could we be justified in trying to change human nature? And, if so, is genetic change an acceptable method? Most of us feel resistance to genetic engineering, and these two questions are often blurred together in our thinking. One aim of the discussion has been to separate the different sources of our resistance. Another has been to try to isolate the justifiable doubts. These have to do with risks of disasters, or with the drawbacks of imposed, centralized decisions. They need not justify total rejection of

> positive engineering. The risks are good reasons for extreme caution. The other drawbacks are good reasons for decentralized decisions, and for resisting positive engineering in authoritarian societies. But these good reasons are quite separable from any opposition in principle to changing human nature.
>
> The idea of "human nature" is a vague one, whose boundaries are not easy to draw. ... Perhaps changes in society will transform our nature. But there is the pessimistic thought that perhaps they will not. Or, if they do, the resulting better people may lose to unreconstructed people in the evolutionary struggle. On either of these pessimistic views, to renounce positive genetic engineering would be to renounce any hope of fundamental improvement in what we are like.[5]

I would say that this reaches the heart of the question. It may be asked what this has to do with limits on the genetic engineering of humans: to which my reply is that, in fact, it has a great deal to do with it.

ON CONSTRUCTING A NEW HUMAN NATURE

If one reads the quote from Glover carefully, it becomes clear that some complex juggling of logic is going on. Following the outlines of the argument, above, about aggressiveness, Glover defends without explicitly saying so the claim that there is a set of behavioral characteristics in humans that constitute a true threat to human nature itself. He then sets up this disjunction: either society (for example, by its educational system) modifies some of these characteristics (e.g., aggressiveness) or it does not. If not, then the path toward disaster has already been chosen. Finally, he adds that, even under the assumption that society *does* manage to modify some of these behaviors, we must face still another question. Why should these results not be improved upon using biological rather than mere educational techniques?

The problem with all this logic juggling is that the whole argument is based on supposition. Assuming today that aggressiveness, xenophobia, etc., are biologically determined – or, to be even more precise, that they are genetically determined – must be recognized exactly for what it is: an assumption in the strict sense, a presumption.

I would not want my terminology to be misunderstood. It is not a "biological foundation" I am referring to, but biological *determination.* The cross-cultural universality of certain characteristic human behaviors, together with the possibility of increasing or decreasing some of them by surgery or pharmaceutical means, proves their biological basis. But people who affirm the appropriateness of using genetic engineering to improve human nature are not limiting themselves to this sort of statement. They are not talking about a biological base but about genetic determination, and they mean that certain genes deterministically cause these characteristics to exist. In these terms, aggressiveness, xenophobia, even sexism are diseases of genetic origin, and they can be cured by appropriate manipulation of the genome, e.g., removing some portion of the DNA string that is responsible for the disease.

Some of my readers might think I am exaggerating; they might say that I should speak of intervention in the genome only if I am referring to diseases in the customary sense – for example, of a possible genetic predisposition to cancer. I do not, of course, mean to exclude that more precise sense. But I would remind readers of what I said earlier about the steam engine – and the same thing is happening today with respect to personal computers and genetic engineering. It is one thing for technical experts to know the limits of their techniques; it is another thing entirely when futurists let their imaginations run wild and end up confusing reality with dazzling possibilities. Nor does it end there; these wild imaginings have an influence on politicians who dole out money for research projects. So it would not be a bad idea – indeed, it would be a good idea – to look out for these extended meanings.

A FIRST STEP IN APPROACHING THE PROBLEM: UNTANGLING THE MESS

Those of us who have taken on the job of evaluating technology find in general that the object of our concern is not always easy to pin down, even in a preliminary way. This is related to the problem I have mentioned, about mixing up promises with realities. This certainly happens in the case of genetic engineering.

To begin with, the terms "genetic engineering" and "biotechnology" are often taken to be synonyms – something that causes endless confusion. Biotechnology usually means a set of techniques (some of them purely theoretical at this point) using living beings in production processes. Some examples of biotechnology practices include the use of ladybugs to combat plant lice, or the use of sterilized females in pest control, or the use of yeast in fermenting wheat or barley. It is sometimes but not always possible to use biological techniques to change physiological, anatomical, and behavioral aspects of living beings, ultimately modifying their phenotype. Both traditional agricultural or livestock-raising practices and modern genetic improvements fall within this category. Genetic engineering, on the other hand, always involves intervention (and the term is uniquely appropriate in this case), delving inside, penetrating to the very material base of the phenotypical features one wishes to modify – namely, the genes.

The two great technological revolutions of our time have this common feature of invading the nucleus. Nuclear engineering invades the nucleus of the atom, and genetic engineering invades the nucleus of the cell. The consequences of penetrating inside the nucleus are in no way similar to the consequences of hydropower production of electricity or the natural generation of species. Saying they are the same would be to ignore, consciously or unconsciously, the real significance of these revolutions.

A second area of confusion is mixing up genetic engineering with the macro-level handling of germ cells. What I have in mind is all those practices that involve the external manipulation of germ cells: freezing of sperm or ova, *in-vitro* fertilization, even extra-uterine fetal development. All of these raise serious ethical issues, but they are very different from genetic engineering – where the techniques involve penetrating the nucleus of the cell, working directly with chromosomes, and inserting, removing, or altering genes.

Having clearly defined what I mean by genetic engineering, I do not want to limit myself to a general assessment or evaluation. Intervening with bacteria in the production of feed for livestock is surely not the same, from a social standpoint, as genetic interventions involving human beings. What I am talking about is assessment not in environmental but in social terms. And within the human domain there are major differences among the genetic engineering of medical products, genetic screening, genetic diagnosis, and gene therapy. Finally, gene therapy that does not affect germ cells is not on the same level as therapy that does, since the latter allows the inheritance of the modified characteristics.

All of these things are often lumped together in discussions of genetic engineering. It sometimes seems as though the use of a microorganism to clean oil wells, the growing of human insulin in E. coli bacteria, using a synthetic protein as a genetic probe, surgery on human subjects to remove oncogenes, and the cloning of humans all generated the same kinds of problems. What we need to note is that, although there are common features, the differences among all these problems, when it comes to assessment, are remarkable. The differences are even greater when the assessment is in terms of threats to society rather than to the environment. Social assessment, moreover, is justified not only because it allows us to plan carefully for the future but also because it allows us, in doing so, to make clear distinctions between designs based on real risks and designs based on technoutopian thinking – that constant companion of technological innovation.

SECOND STEP IN APPROACHING THE PROBLEM: A SOCIAL ASSESSMENT OF GENETIC ENGINEERING

Technologies can be assessed in terms of impacts or risks. I have elsewhere dealt with the assessment of the impacts of genetic engineering,[6] and though I could repeat some of that discussion here I want to note one thing. When one speaks of manipulating the human genome, little can be said about impacts; in the domain of human genetics, what we are dealing with is as often carpentry as true engineering. Moreover, it is important in this context to specify whether one means to use environmental standards (as proposed by Glover; see above) or social standards.

I think everyone is aware that technology assessment has, in an incredibly short period, passed from non-existence to environmental impact assessments to social assessments. Moreover, it seems equally clear that environmental impact assessments based on economic criteria are, in recent times in developed countries, in a state of crisis. The crisis arose precisely at the same time as the period of new affluence. Under these circumstances, in some places, exactly the same value is attached to the environmental impacts of particular technologies as to the social changes their introduction could bring about in terms of what might be called "the way of life" of those societies. Here I mean the forms of life regulated by values, folkways, customs, economic and social relationships, etc.

I want to cite a simple example. An industrially produced protein with a structure similar to that of human protein would have foreseeable effects not only on the environment but also on the emergence of life forms suitable to the level of danger of the product. A similar and even clearer example is that of the large-scale production of plutonium.

The truth of the matter is that, whereas environmental assessments can generally be reduced to specific monetary figures, social assessments generally involve factors which are difficult if not impossible to reduce to economic terms. We can not ask how much it is worth to reject certain values or moral principles. It may, it is true, be possible to make certain predictions about relative advantages and disadvantages of introducing or not introducing some particular technology. But making these comparisons in terms of "ways of life" would inevitably introduce subjective factors into the evaluation. And if anything is clear today, all things considered, it is that social factors must be taken into account in the assessment of proposed technological developments.

SOME CONCLUSIONS

1. The question of limiting the use of a particular technology – in this case, genetic engineering applied to human beings – must depend on both environmental and social assessments of its introduction.

2. Social assessment, in such cases, must include the evaluation of both impacts and risks; and the paradigm for such complex evaluations should be the anticipation and avoidance of future problems, not just the solution of current problems that may have been unforeseen. It is difficult to make such predictions, but at least today we have sophisticated computer programs to help us out.

3. Social assessments need to be both generic and specific. Generic assessments of particular technological developments – such as the one being done here – reflect often confusing aspects and concepts of *Weltanschauungen*, visions of the way the world *ought* to be at a given moment in the history of humankind.

4. At a more specific level, it is claimed that genetic engineering can modify human nature in substantial ways. But what "human nature" means here is very difficult to say. What is clear is that advocates for the improvement of human nature by means of genetic engineering make their proposals on the basis of certain assumptions. These assumptions are by no means naive, and they could involve sweeping changes in social forms.

In effect, these advocates of human genetic engineering are making the assumption that the most salient features of human social behavior are genetic in origin. Furthermore, they are assuming that the genetic base causes these traits (which might generally be labeled "innate") in the strict sense. They are assuming that the most salient features of human social behavior are phenotypically unavoidable; they will manifest themselves whether we want them to or not. Moreover, they are assuming that social attempts to deal with these forms of behavior, attempting to channel or inhibit them, have little or no effect. And finally in many cases they are assuming that these genetically-based behavior patterns will not lead to salvation but to slaughter. It is on the basis of these assumptions that they end up justifying first the possibility, then the necessity, of applying genetic engineering techniques to make human beings more human. (This last, of course, is a little play on words.)

In this context, it becomes clear that our earlier generic evaluation makes good sense. Its validity is based on two considerations. First, it untangles us from a sticky web of often bold and speculative assumptions that support the recommendation for scientific and technological intervention. We can see that what has been presented as an objective, neutral, more-or-less "true" recommendation that accords with scientific practice may well belong, instead, to the realm of fantasy.

Second, the generic assessment is interesting because it also shows, in pointing out the speculative nature of the nest of assumptions that there really is no scientific basis for the particular social experiment – applying genetic engineering – that is being recommended.

A simple example will help clarify this. The United States, early in this century, in its immigration policies, separated Slavic and Latin immigrants from North European immigrants, claiming that Eastern and Southern Europeans showed high rates of mental retardation. This was based on tests administered to immigrants on arrival, and it reveals a belief that severe mental retardation is biologically determined. This same belief, alleged to be scientific justification, also undergirded laws restricting immigration from Eastern and Southern Europe. Furthermore, the belief was used as a standard to inflict sterilization on those who scored low on the tests. All of this was based on *assumptions*, tied to a specific technique for measuring psychological deficiencies, that later came to be recognized as ludicrous. (At least they seem ludicrous to those of us who have studied them.)

4. It is important – after clearly distinguishing, in proposed applications of genetic engineering to humans, what belongs to human intervention in general from the specifics of macroscopic treatment of germ cells, pharmaceutical production, genetic diagnosis, and gene therapy – to avoid vagueness, to go directly to specific environmental and social assessments of each of these technological applications.

Keeping this approach in mind, I want to say again that limiting genetic engineering means something very different in each of these cases. Using genetic engineering to manufacture pharmaceuticals, for example, does involve environmental risks, including some that may affect human health. Nonetheless, assessing such risks is a matter very different from assessing techniques of gene therapy.

Concretely, what I have in mind is this. Although there are common features in both sorts of assessment, this should not blind us to significant differences. One example of a common feature is that both techniques are claimed to be both clean and profitable. Genetic engineers also commonly defend their techniques as providing benefits to humankind. If, now, we look at the situation without blinders, it seems clear that, since the beginnings of genetic engineering, profits have been much more important than benefits to humankind. I do not deny the importance of profit, but it would be naive to create an image of selfless crusaders struggling to cure the ills of humankind.

One example is genetically-engineered pharmaceuticals, where the claims are the exact opposite of the reality. At one point it was claimed that, with special bacteria, it would be possible to produce human proteins which, if defective or missing, caused certain diseases; and they could be produced at ridiculously low cost and with no pollution. The genetically-engineered wonder drugs – insulin, somatostatin, interferon – were to be the hope of the future. Assessors have judged the results to be mediocre, even in economic terms; production alone is not enough and commercialization jacked up the prices. The idea in the abstract is one thing, but it is something entirely different to put a technology into practice in a given socioeconomic system.

Another example is the genetic engineering of vaccines. Again it was claimed that a cheap means of producing them had been found, offering the hope of eradicating common diseases especially in underdeveloped countries. But again reality presented a problem: these countries had no money to buy the products – which has led to the abandonment of these proposals.

In short, an assessment that would set limits on a particular technological development can not be based exclusively on the claims – on the special *Weltanschauungen* – of the practitioners of the technology in question. Neither can such an assessment be based on promises. It must take into account, above all else, the actual results; from these and certain other parameters, reliable predictions can be made.

In the example above, it was clearly foreseeable that prices would be higher than predicted. Again, we have to recall that it was human proteins they were synthesizing, so it was inescapable that there would be consequences and risks for employees in the factories where they were produced. (Similar consequences face employees in factories where non-human proteins are developed.) Introducing measures to control these problems – along with numerous other problems associated with the sophisticated level of the establishments, the type of bacteria used, etc. – would eventually lead to the conclusion that what had seemed *the* profitable business of the late twentieth century was not profitable at all.

I want to repeat here what I said before: I am not opposed to doing business for a profit. But assessments must be based on reality, not fancy.

In making what I have been calling generic assessments, it is clear that we can turn up some notions that are mere fancy. But in making specific assessments – recall the example of the genetic pharmaceutical industry – we are forced to keep fanciful notions from interfering with assessments either in environmental or in human terms. The promises associated with

Weltanschauungen must be assessed in generic terms, but then it is time to get down to specifics and assess the actual and potential consequences of particular technologies. Claims that something is going to eradicate disease in the Third World, or that genetically-manufactured vaccines are going to prevent the negative consequences of using non-human proteins, or that these same vaccines will alleviate problems created by missing or defective proteins – all these claims can be assessed without referring to fancies that obstruct clear vision of problems.

From where I stand, I think I see signs of this sort of realistic assessment. Pharmaceutical manufacturers now recognize that results have not matched high hopes – though they may still hope things will change in the future.

Such realistic limits are what we should expect with an activity that involves synthesizing human products. Environmental risks are obvious – though, in principle, there are measures available to control both the bacteria used in the process and the products of the system. Though these measures are costly, they are essential. Problems arise when genetic researchers turned businessmen do not recognize the risks – as happened at the end of the 1970s – because they are deluded by promises more in line with cosmic visions than with down-to-earth production.

It is salutary here to recall how – when confronted with guidelines proposed by the National Institutes of Health but in fact written by the genetic researchers themselves – the genetic engineers moaned that these guidelines were an impediment to human progress. (James Watson is a good example.) But this kind of talk betrays what it is, mere speculation superadded to a technological venture that ought to be assessed on its own merits. That is what the matter is all about.

ADDENDUM: THE HUMAN GENOME PROJECT

I come at last to the most important issue, and here again distinctions are called for. When a practitioner of genetic engineering, or a futurist discussing the subject, writes something on the advantages of doing gene therapy on human beings, the article or book typically ends up lumping together very different conceptions of human nature, genetic production of pharmaceuticals, the likelihood of incarceration of genetic defectives, the creation of genetic ID cards, and of course – that was the main point – genetic therapies. It is by referring to these virtues that the author manages to suggest there are other virtues in extending the techniques to other domains. A typical author begins by citing the more than two thousand diseases of genetic origin that have been discovered, then goes on to say that by replacing, repairing, or eliminating certain genes these diseases could be eradicated. Such an author then typically adds that the process could be even more effective and beneficial if the patient did not, if possible, have to ingest chemicals or products of biosynthetic origin. (Here there is one clear benefit, if the patient does not need to take medications – which is something positive!)

Now just as the creation of new genetic ID cards needs to be assessed in both generic and specific terms, the same is the case with gene therapy. It is necessary to avoid (once again) that all-too-frequent ploy of the futurist defending genetic engineering – the ploy of starting to talk about gene therapy, then ending up talking about "constructing a new being." If this is going to be done, it, like gene therapy itself, needs to be discussed in its own context and on its own merits; it should not serve as a mask – and in this case a harmful mask! – for the gene therapy with which it is linked.

To move the discussion toward a conclusion, I want to add two more points.

1. Although, as we shall see the term "gene therapy" covers a variety of techniques, all of them do share one common feature – a feature they share, in fact, with other forms of genetic engineering: namely, they all appear to be good business ventures. I would not (once again) claim that financial gain is necessarily bad, nor that there are no humanitarian ideals involved. But it is hard to deny that, in ventures such as the famous genome project, financial concerns are involved – to deny, in fact, that they may be crucial. Clearly learning the sequences of genes in the human genome could be the basis for eradicating some quite serious diseases, but equally clearly it will produce large profits for the corporation that undertakes the venture – particularly if it leads to some sort of patent. This is undeniable, and it helps to explain the race that is on between the Japanese, American corporations, and European institutions. I realize, of course, that it can be uncomfortable to call things by their right names; it can even put a person in a compromised position. But a process of assessment that would analyze all impacts, make predictions about clearly foreseeable risks, and set limits on a technological venture must do exactly that.

2. The techniques for various forms of gene therapy are so different from one another that a methodologically sound process of assessment requires a different assessment for each one. To be specific, an assessment of a gene therapy that involves no negative consequences for the patient *must* be different from an assessment of some other therapy that, for instance, either interferes with germ cells or involves the cloning of humans (or even of some

limited human characteristics). The first sort of therapy is clearly similar to current practices in medicine, and is subject to the same sorts of norms that are operative there. Responsibilities would be adjudged in accord with extant codes of medical ethics. With the other sorts of therapy we must invoke such vague concepts as "human nature" since we may well be having an impact on organic structures that are the basis for certain characteristic human behaviors.

Though it may seem so, I am not here contradicting what I said earlier. I am not talking about particular genes as determining particular features of human behavior; what I am saying is that, although not determined by particular genes in the strict sense, some characteristics of human behavior do have a biological base, and interference or inserting synthetic genes might alter certain behavior patterns by altering the biological bases on which they rest.

Here is an example: although there is no gene for human aggressiveness, organic structures such as the amygdala are related to behavioral manifestations of aggressiveness. One way of getting rid of aggressive feelings in ourselves is to take a tranquilizer; another way is to have an area of the brain surgically removed; and still another might be to interfere with the genetic material that underlies the synthesizing of proteins that constitute the organic base of the behavior in question.

This last possibility is the most difficult to assess. One reason is that it is so closely linked to the *Weltanschauungen* of those making grand prognostications about the future of genetic engineering – to return to what I said at the outset. Obviously, in such a case, limit-setting involves not only environmental and social standards but also philosophical principles and ethical ideals. This must necessarily be so because such prognosticators are talking about a future that does not belong to us but to our descendants. Setting aside questions of technical feasibility (at least under current technical conditions), what I am emphasizing, under the heading of risk avoidance, is that we must ask ourselves whether we are justified in making synthetic selves. However one answers this difficult question, one thing seems clear now. The matter cannot be left exclusively in the hands of the practitioners of genetic engineering, nor in the hands of their futurist friends. We need, to repeat, social assessments of cosmic visions and concrete, specific assessments of each technique individually and realistically.

Polytechnic University of Valencia

NOTES

* This paper was prepared with the help of a grant (PS87–0128) from the Spanish Ministry of Science and Education (CICYT). I want to thank the editor of the volume for help on the English translation; members of the Science, Technology, and Society program at Pennsylvania State University provided helpful observations on the original version.

1 See my article, "Tecnología democrática frente a democracías tecnológicas," *Arbor*, forthcoming.

2 See my article, "Reflexiones en torno a la cuestionable primacía de lo teórico, o semblanza del cachivache," *Arbor*, no. **507** (March 1988):29–45.

3 I have in mind the books of Konrad Lorenz and Desmond Morris – *any* of their books.

4 See my article, "El 'Pequeño' Paso que media entre ensenar los dientes y enviar misiles," in J. Sanmartín *et al.*, *Genética y conducta* (Valencia: Tirant lo Blanch, 1986).

5 Jonathan Glover, *What Sort of People Should There Be?* (Harmondsworth, England: Penguin, 1984), pp. 55–56.

6 See my book, *Los Nuevos Redentores* (Barcelona: Anthropos, 1987).

RAPHAEL SASSOWER

THE ALARMIST VIEW OF TECHNOLOGY

I

Individuals and institutions can adopt either a collective technophobia and attempt to ignore or to minimize their use of technology, or they can adopt a technophilia and enthusiastically maximize their dependence on and support of technology without any concern for important risk factors.[1] Between these two extreme responses to technology there is a third possibility that attempts to achieve some equilibrium, accounting for the problems associated with technology without thereby condemning technology as a whole. The third option expects to optimize the advantages technology has to offer, while keeping under control most of the potential drawbacks associated with technology.[2]

In what follows I provide a partial illustration of what I call the alarmist view of technology. This view seems counterproductive: while assuming that alarm is necessary for counterbalancing the enthusiasm associated with technophilia, it has scared away many people from consideration of the problems associated with a response to the pressing issues that arise from the deployment of modern technology. Put differently, the goal of the alarmist view is to alarm people, press upon them that complacency will not do, and that one should never minimize or ignore the hazards accompanying what is termed modern high technology. However, such a goal may not be achieved through the means the alarmists use, since their *techniques* (and not their general goal) are counterproductive. The techniques of alarm, in whatever manifestations, are more likely to frighten people and not invite them to join in the debate about the deployment of modern high technology. And individuals who are suspicious, if not outright frightened by technology, might back away entirely from the debate and will not lend their support to those committed to maintain some balance between fear and hope *vis-à-vis* technology.

It is presumed that a balanced opinion is far more appropriate than an extreme one. This holds true, so it seems, whenever there is a controversy or a debate in which there are at least two sides, each of which is able to master

Paul T. Durbin (ed.), Broad and Narrow Interpretations of Philosophy of Technology, 167–180.

some good arguments in its support or defense. This, of course, is not the case when one observes an obviously one-sided issue, such as the extermination of Jews in gas chambers (medium and not high technology) during World War II. In face of such an extreme situation, one cannot expect that the annihilation of the Jews can be defended with the same weight of the arguments that can be constructed against such a travesty.

This essay is not concerned with technology itself, its various manifestations, its numerous practical impacts, or its assessment. Instead, it is concerned with the debates about technology and its various modes of assessment. And more specifically, it is focused on the techniques employed in the debates. Finally, the emphasis is on discarding one particular technique, that of alarm, that overshadows all other techniques in its power of expression and in its appearance of urgency. That is to say, when such a forceful technique is used, it has the power to eliminate the possibility that other techniques – and through them other perspectives – will join in the debates about technology (low, medium, or high).

II

Sounding the alarm is neither beneficial nor harmful in itself.[3] We easily distinguish between sounding the alarm in order to save the lives of people who may not know that their house is on fire and sounding the alarm for fun in a crowded theater. This distinction is upheld by courts of law, at least in the sense that one's freedom of speech can be limited only in the latter case. But there are some difficulties in distinguishing the less obvious cases; difficulties which result in legal complications as well as in complications that surpass the legal domain. For example, is it enough for one to sound the alarm only, or must one enter the burning house or dive into the frozen lake to save lives? Questions like these are of no particular interest in the present context, for it is unimportant what is the legal status of the use of alarm. There is no legal injunction against the use of alarm in debates. However, just as in the case of a crowded theater in which an alarm is sounded and some people are hurt for no reason (since it is a false alarm), alarmist techniques have practical consequences that can be deemed harmful.

What is so interesting about discussing the sounding of alarms? At the very least, two things. First, alarmists may in effect prevent potential participants in the debate over technology from joining and contributing to finding new means for handling the difficulties associated with the application of technology. Second, consider the classic story of the shepherd boy who cried wolf too many times and was finally ignored.[4] The goal of alarmists is not to shut off the debate, but presumably to highlight certain problems that technology brings about. However, alarmists may be ignored if they continue to call out falsely, that is, to predict dire consequences that have not yet materialized (of course, in part due to their appropriate warnings).

The focus on the technique of alarm (and not on the goals that can be achieved when sounding the alarm) is twofold. On the one hand, it is a focus on the techniques used by those sounding the alarm, the alarmists, the extremists in the debate about technology. On the other hand, it is also a focus on learning *how* to listen to and interpret alarms. This dual focus may be more effective than either trying to curtail the impact of the alarmists or attempting to educate those privy to the debates about technology. For example, by learning the techniques of listening, and thereby learning about the techniques of expression, we may learn to distinguish between what could be interpreted as frivolous cries and those worthy of attention. Once such a distinction is made, we can prepare ourselves to calm what may be considered a paranoid crying person or save the life of what may be perceived as a threatened person.

In some respect, then, it may be argued that it is more fruitful to concentrate public attention on the debates about technology and not the proliferation of this or that specific technology. It seems that contemporary life, as Albert Borgmann agrees, is technological, and inescapably so.[5] This may mean that for all practical purposes there is no longer the question if we should or should not have technology, but rather how much of it is acceptable or necessary – what kind of technology is more appropriate and to what specific uses should this or that technology be geared. And this practical condition necessitates a shift to another level of public concentration and orientation, a meta-level of technology, the level of the debate about technology.

From what was said so far, it should be clear that the meta-level is indeed a technology of technology, a meta-technology. That is, at the meta-level there are various techniques that can be used, techniques of debate. So, if one wishes to have an impact on contemporary life, one may do so more effectively through the invention or revision of one or more techniques of debate than by climbing over nuclear reactor fences or burning nuclear armaments. Even though participants in the debate presume that they have something to say to the general public and thereby affect public policy, their audience is usually much more limited; it includes other scholars who share their jargon and entrenched terminology to the exclusion of the general public. In this respect, then, the concern with the techniques of alarm is a concern about the discouragement of the general public from participating in whatever form in the debate about technology, so that the public will shape the future of its technological life as opposed to having it shaped by a small intellectual minority.

III

The folktale of the shepherd boy's alarm reappears in a different technological guise, with modern technology in the modern world, than the one commonly understood some fifty or one hundred years ago. The wolf is modern technology, a beast that threatens to devour everyone. How exactly one approaches this beast, whether it is a loner or an animal living in a herd, whether it can be tamed or not, remains an open question. Jacques Ellul, for one, spends some four hundred pages in *The Technological Society* attempting to develop a preliminary answer to this question. He distinguishes between technology and machines, between *technology* or *technique* and the industrial apparatus, and between the planned use of means to achieve predetermined ends and the autonomy of technology, where means become ends. Ellul's lengthy discussion refocuses one's naive view of the concept and proliferation of technology and helps shape a more sophisticated understanding of the actual instruments associated with technological innovations and the human capacities and characteristics – from reason to consciousness – as they pertain to the technological phenomenon.[6] Since his answers and definitions are commonplace, they are echoed around the world. They serve to some extent to perpetuate myths about the wolf.[7]

Myths can serve as illuminating, educational tools; but they can also be used to establish taboos or to induce society to avoid certain issues. Of course, it is possible that the myths of technology have had effects contrary to those hoped for by their modern inventors. Some may argue that instead of educating and forcing contemporary society to think beyond a level of technophilia, a level of fascination devoid of any critical reflection, the myth of and about technology has evolved to a level of technophobia, a level characterized by fear and paralysis. It seems that by now either of these two extreme positions turns out to be ineffective in light of the realization that there is no one single way of defining technology or separating particular technological developments from the social and political context in which they evolve.

Besides, as mentioned above, the focus must be on the process by and through which society achieves some agreement concerning its policies towards science and technology. That process, though seemingly technical in nature – where technical data and jargon are routinely used to explain and defend certain positions – can be interpreted as being politically motivated and socially affected. Since what is called the alarmist view in this context is closely aligned with one of the extremes, the extreme that warns society against the evils of technology or that technology is evil, it becomes a form of oversimplification of a complex situation that orients the debate in one direction alone. Such orientation may do more harm than good to a society that is attempting to learn how to cope with technology. Even though the alarmist position may be viewed as a way of coping, coping here is not limited to the ability to interact with specific technologies, such as nuclear power or cars or computers; instead, it incorporates also the ability to participate in a somewhat intelligent way in the discourses, the debates, the setting of public policies, and the establishment of particular local guidelines.

Of course one can argue that the counterproductive or contrary effects mentioned here are exactly what alarmists are interested in achieving. That is to say, alarmists hope to achieve a moment of critical reflection or what some may call paralysis so that their messages will be heard around the world in order that what are considered the most dangerous or inappropriate technologies will be arrested. In this sense, then, the alarmists may be perceived as contributing to rather than undermining public participation in the debates about technology. However, their contribution can be simultaneously perceived as suspect, since it may be undertaken without the consent of the public. In this light, one can still sympathize with the long-term goals of the alarmists and at the same time point out a strategic blunder.

Strategically, one can argue, the alarmists are not concerned enough about the techniques they use and to what extent they are crucial in setting the stage as well as performing the activities associated with the debates about technology. And this point, as mentioned above, warrants criticism.

Instead of diverting or stopping the engines of technological progress even for a moment, all alarmists can hope for is influencing or stopping only the debates about technology and technology itself. But this point does not imply that alarmists are ineffective or unimportant. If that were the case, there would be no reason to criticize their techniques. Since, because their message can be perceived as clear-cut, they may speak more clearly to less informed people than to those in control of public policies, their overall impact and appeal may be greater than expected. Instead of being marginalized by their radicalism, they may become the heralded spokespeople of the general public.

The debates about technology, technology's current state or states and future prospects, are important in shaping social attitudes towards technology in terms of public funding and general acceptance. Alarmists, then, are in a position to have an impact on how public policy concerning science and technology is shaped. To some extent, they are using the democratic structure to their own exclusive advantage, rather than using the democratic structure democratically, leaving the debate open to any other extreme or moderate viewpoint. In this respect, then, alarmists, as extremists, use means and techniques that eliminate the possibility of an increased number of participants in debates where the sheer size of participation is most important because the size of participation represents a democratic commitment.

When debates turn into shouting matches – because the participants have fixed ideas concerning technology and its effects – when they are exclusive in the sense of not allowing differences to reshape the debate or when they use techniques that exclude additional information or opinion from surfacing, it is difficult to find out what issues they attempt to address or if the same issues might be addressed differently. Do they deal with specific technologies or techniques, particular instruments and machines, or with the cultural context that gives rise to certain ways by which technologies are developed and deployed? One may object that the concern of the critics of alarmism is based on a confused definition of technology,[8] a confusion already flagged by Ellul. Most alarmists' warnings have focused on particular technologies or techniques – artifacts or types of artifacts, such as nuclear power plants, chemical and biological weapons, and genetic engineering techniques. Is the critique of the alarmist position limited to the treatment of these sorts of artifacts, or is it extended to the broader sense in which technology is understood – as forms of consciousness rooted in particular social relations and/or social institutional dynamics?

The image of the wolf can easily accommodate both the narrow and the broad definitions of the concept of technology: it is an *ensemble* (to use Ellul's term) of animated techniques and artifacts. But it is more than an ensemble as well; it is a creature whose thought processes, however defined, are not bound by any preconceived human framework, because it is an *animal*. An "animal," as the *Oxford English Dictionary* tells us, is an "organized" being that is "endowed with life." When used "contemptuously or humorously," animal refers to a human being who is considered "no better than a brute, or whose animal nature has the ascendancy over his reason."

Such a brute, to be sure, is also characterized by Ellul as an independent creature whose independence renders its behavior unpredictable.

To some extent, that which is animal-like is both an organized being full of sensations and at the same time not organized enough to follow a certain rationale, the reason and reasonableness of human beings. Is this a liability or an asset? Should the wolf be celebrated or persecuted? This last question is especially troubling since one needs to assume a certain classification and categorization of what technology is before one can answer such a question. In other words, in order to justify sounding an alarm about the wolf, one needs to be more specific about what exactly is troublesome enough to warrant such alarm. This proposition is not an alarm about the alarmists, since there are many occasions where sounding the alarm can be widely interpreted as an appropriate mode of reaction.

For purposes of the present discussion one need not assume either that there is a set criterion by which appropriate technologies can be distinguished from inappropriate ones or that one set of suggested criteria is superior to another. Any decision about the appropriateness of technologies remains context bound in the sense already developed by pragmatists. That is to say, one should not predetermine the positive or negative aspects of a particular technology until the consequences of its deployment are made clear; and of course such clarification comes about only after the fact, allowing only retrospective rationalization processes that help accuse or vindicate a specific position one has already adopted.[9] Regardless of the differences in the perceptions, in the assessments of the risks and benefits of specific technologies, that alarmists uphold, the present critique considers them all to be guilty of using dangerous tools.

IV

It is possible to begin the partial chronology of the alarmist position with the Industrial Revolution, recalling the outraged reaction of the Luddites who vented their anger through the destruction of machinery that was produced during those revolutionary years. The present account begins with the works of Max Weber and Thorstein Veblen. They represent some of the concerns voiced in the beginning of the twentieth century both in Europe and in the United States. It should be noted here that I consider Weber and Veblen, Martin Heidegger, Herbert Marcuse, and Ellul, and finally Hans Jonas all to be representatives of the alarmist view of technology. The list of alarmists is not exhausted with these six; their voices are the most pronounced and well known.

Both Weber and Veblen recognize, in their own ways, how technologically oriented modern society has become. In Weberian terms, that means primarily an understanding of the bureaucratic structure that dominates our lives; a technical structure so rational that it lacks human passion and aspiration; a structure so powerful in its techniques that it may devour our very souls.[10] In a similar fashion, Veblen warns us against the technicians who are quite capable of revolutionizing our country in the Soviet sense. When he was writing in 1919, the Bolshevik Revolution was still vivid enough to inspire him to sound the alarm as loudly as he could.[11]

It is true that Weber's description and warning differ from Veblen's. Weber argues that we are all doomed to become bureaucrats, wolves, or thieves. By contrast, Veblen argues that we ought to watch out for the wolves or thieves among us – the technicians. It is difficult to determine whether Veblen takes technicians to be a threat or of great emancipatory power. Even when he condones them for their ability to transform the American landscape, it is unclear what mode that transformation will adopt.[12] Technicians, according to Veblen, are necessary to fight the two evils of modern American capitalism of inefficiency and absentee ownership. But will the world they help bring forth be a better one? Will their techniques indeed save the world Veblen wishes to save, or will their techniques be superseded by other capitalist techniques worse than those used before?

Regardless of their differences, the threat of technology and technicians is nevertheless imminent for both Weber and Veblen. As social critics they feel it is their responsibility to sound an alarm, warning their fellow citizens against the potential hazards of technological innovations and the overly rational or bureaucratic mindset that accompanies these developments. In this case, then, the concern is not with specific things, such as locomotives or bridges, but with a way of thinking, a technological orientation and attitude.

Some might say that the twentieth century presents an exaggerated extension of the nineteenth-century romantic view of technology. In this respect, the view of technology has become more confused, even incoherent, from the First World War through the end of the Second World War. Technology is an evil, a destructive force that kills millions; yet it is also a necessary evil, something modern society cannot do without in order to protect itself from invaders, defend itself, and attempt to ensure its very survival. Technology in the sense of specific techniques, such as medieval instruments or efficient modes of food production, is no longer the only technology that the public is confronted with. By the beginning of the twentieth century, technology is understood as pervasive and intrusive – it has become unavoidable. If technology cannot be shut off, can it be controlled? This question was difficult to answer especially in light of the use of war-related technologies, where bystanders inadvertently suffered from the consequences of events – wars or the change of modes of production – from which they did not directly benefit and to which they had never given consent.[13]

It seems that from that time on, the alarm sounded by the works of Heidegger, Marcuse, and Ellul is of two kinds, global and local – but not in a geographical sense. Global in the sense of studies in the ideology of the industrial society, which is, incidentally, the subtitle of Marcuse's *One-Dimensional Man*. Global in the sense that technology must be recognized, with Weber and Veblen, as the foundation on which modern society makes its one-sided progress, a progress towards that rationalization of every aspect of life, a revelation of whatever truth it is seeking to uncover, and, thus, a reduction of the complexity and beauty of life to a set of scientific methods and technical rules. One may call this a global alarm because this view advocates a revolution in thought as well as a radical change in the social and political and economic situation which we have created.

These works are critical in nature; they criticize a way of thinking or doing, so that a better approach to and interaction with technology can come about. So the critique of their critiques is not a critique that disagrees with their views nor does it disagree with their perceptions of the situations they describe. Instead, it is a critique of the methods they use, the means and techniques with which they confront their audiences, and by which they frame the debate about technology.

While Marcuse focuses on what Heidegger terms the "potentially lethal machines and apparatus of technology" as employed in the hands of capitalists for the purpose of increasing exploitation and alienation, Heidegger is concerned with the danger associated with technology that bars the possibility of revealing "a more primal truth."[14] Heidegger announces the threat of technology in the sense of the loss of human control. Some may view the pronunciations of this global view as either phenomenologist or Marxist in nature, but these labels add little to the understanding of the issues these authors are concerned with because these labels carry with them too broad a philosophical spectrum and too narrow a political objective. It may be best to replace the phenomenologist/Marxist label with the global one proposed here.

The alarm sounded by Ellul in *The Technological Society* is both global in the sense of being concerned with human consciousness and local in the sense that it is focused on technology itself.[15] Part of the critique is localized by placing technology at the root of all the troubles facing contemporary society. There is a parallel between the works of Marcuse and Ellul; both focus on the same culprit and then turn to what they consider the devastating future that lies ahead. But while Marcuse goes beyond the technology faced by modern society to express the sad state of human intellectual activity in a Heideggerian sense, Ellul is more concerned with an internal critique of technology itself. For Ellul, the very term "technology" is puzzling enough, or too ambiguous, to be left untouched. Heidegger recognized an ambiguity, too, but for him this ambiguity prompts a discussion on the nature of revealing truth. And Marcuse is concerned with defining technology in ideological terms.

Whether the ultimate concern is with truth or ideology, whether the concern is to reform society or change its social structure and ways of thinking, in all cases no bystander could ignore the alarm any longer, for ignoring it becomes a deliberate act. In this respect, the focus should not be on participants alone, whether they are alarmists or not, extremists or not, but on bystanders, those who are potentially participants but who have been discouraged so far from participating. The question is therefore a question of how does one move bystanders to take a stand, to become involved. So, are bystanders deliberately avoiding the debates about technology or are they discouraged from joining them? From the perspective of the critique of the alarmist view, and especially its twentieth-century manifestations, one may argue that bystanders are not encouraged to join the debate.

But is there no difference between Heidegger, Ellul, and Marcuse in the respective ways they sound an alarm? A few lines from each work may illustrate the difference that may still distinguish them. It seems that while Ellul's work is alarmist in the sense of leading to paralysis, Marcuse's and Heidegger's are critical in Veblen's sense; that is to say, though alarmist in nature, they initiate a response. Marcuse has more precise political, social, and economic objectives, but Heidegger too encourages continued action, even if it can be seen as too intellectual. In this sense, then, a global alarm may become eventually more effective than a local one. This is counterintuitive, for we are used to believing that responding to a local alarm leads to noticeable results, while global alarms are too remote and overwhelming to lead to any action at all. For example, we think a community is more likely to respond to an alarm about the contamination of its water resources than to an alarm about the Brazilian rain forests some three thousand miles away.

Marcuse begins *One-Dimensional Man* by discussing "The Paralysis of

Criticism," and encourages society to develop some viable opposition to the trend in which it finds itself.[16] His concluding remarks also emphasize the alternatives to the one-dimensional society, alternatives that he considers starting points from which to transform the devastating results to which modern industrialization may lead. Marcuse's critical description contains a message of hope, at least in the sense of a critique which may change our thought, actions, and lives. His view, in a manner analogous to Heidegger's, suggests a critical response and is not fatalistic. His view seems different enough from Ellul's not to warrant the criticism which Ellul's view may deserve.

Though Ellul admits to not knowing what course the "disease" of technology may follow – that is, what will be the results of technological innovations – he still feels like the "physician" who must expect the worst – death from his prognosis. Ellul calls this a diagnosis.[17] And there is every reason to believe that the course of the disease will turn out to kill the patient, because Ellul believes that the disease, technology, has become autonomous.[18] There is a sense in which, to continue Ellul's medical metaphor, the disease in question behaves like a cancer which multiplies its cells until it kills the body that nourishes its growth. The physician may hope for miracles, but it is most likely that the patient will die. Just like the helpless patient facing cancer, Ellul describes the cancer of technique in the following terms: "Technique is essentially independent of the human being, who finds himself naked and disarmed before it."[19] This view of technology is alarmist at least insofar as it allows modern society to view itself as impotent to do anything about the negative aspects of technology. It is defeatist, even if it pretends to provide description of something other than the conditions of human paralysis.

Ellul's view, however, only paves the way for the most prominent alarmist view of technology – as a barrier to further debate, as a means of stunning people from entering the debate – that presented by Hans Jonas in *The Imperative of Responsibility*.

After listening to Jonas, for example, society is more likely to suppress anything that has to do with technology from our consciousness, rather than attempt to understand and appreciate the complexity associated with debates about the technological society. Leave us alone, society may beg, we cannot even comprehend what is at stake, let alone imagine what our role ought to be. Better to get away from the whole story, close the book, and pretend it never happened. Besides, what voice do we have? What power do we have to reject or endorse the views of alarmists?

The main contribution of Jonas to the degeneration of the alarmist view is that he is most explicit in his use of alarmist techniques. His exaggeration is so blunt that it needs some reflection. That some fear of technology, like a fear of anything capable of killing, needs to be incorporated into the social decision-making mechanism is quite reasonable;[20] but that fear alone should be the guide is both unreasonable and paralyzing. For example, Jonas requires us as members of modern society to be "obliged to lay ourselves open to the appropriate fear,"[21] a fear he does not wish to be thought of in remote and future terms, but in quite realistic and immediate ones.

It is also quite reasonable for Jonas to point out that the sort of technology society is facing today is radically different from the one faced by his predecessors, like Weber, Veblen, Heidegger, Marcuse, and Ellul. Contemporary technology is changing the environment in which we live, thereby replacing the foundations we took for granted with a new set of conditions unknown until fairly recently.[22] Yet, when Jonas belittles the concept of hope, which he associates with utopian thinking,[23] Marxist or other,[24] he is definitely moving into one extreme of a spectrum of views and positions. Even though he concedes that he may be charged with "one-sided emphasis," even perceived as holding an extreme anti-technology view, he defends his pursuit in light of the fact that he is writing "in a time of one-sided pressures and mounting risks," so that his own words may be on the "side of moderation and circumspection."[25] Regardless whether his view may be needed temporarily, as he believes, or for eternity, the presentation, its techniques, may block potential alternatives and the further expansion of the debate about technology.

V

The public might be frightened by the techniques that Jonas and other alarmists are using in the debate about technology and thereby avoid any sort of participation. A critical eye, maybe even a concerned eye, must be cast by the public towards technology. An eye, though, which can be reflective as well as involved enough to appreciate how important a role technology is already playing in our life, and how we ought to interact with it through public policy and public debates. And here institutions may be of help, especially if they are viewed as democratic and non-static. For example, the public may wish to curtail the ability of government agencies to collect and cross-reference data bases concerning individual citizens so that the privacy of individuals, as a right, will not be compromised. Further, the public may wish to change the legal code concerning private property in regard to computerized data, at least in the sense of who ought to have access to that data.

In this sense, then, technology itself is a dynamic institution; it can be democratically run and the public educated on the question of how much alarm is conducive for action and how much is paralyzing. Instead of reacting to the latest cry of alarm, the public should view alarms as part of a larger debate, as a necessary means to wake it up from its dogmatic slumber, as Kant said of Hume, but not too much to make it want to ignore or close the debate altogether.

University of Colorado at Colorado Springs

NOTES

[1] I wish to thank Gayle L. Ormiston, Charla Ogaz, and two anonymous referees for their critical suggestions on earlier drafts of this paper. The terms "technophilia" and "technophobia" are discussed in Alan R. Drengson, "Four Philosophies of Technology," in L. Hickman, ed., *Philosophy, Technology, and Human Affairs* (College Station, Tex.: Ibis Press, 1985).

[2] The notion of technically balancing the extreme views concerning technology is best illustrated in Joseph Agassi, *Technology: Philosophical and Social Aspects* (Dordrecht: Reidel, 1985).

[3] It should be noted that the terms "sounding the alarm," or "sounding an alarm," are used interchangeably here, since there is no clear and unique alarm that dominates the present discussion. In addition, the term "alarm" is here meant in the broadest sense, since the definition itself is broadening daily due to an increasing number of technological innovations that allow the alarm to be sounded and heard beyond and regardless of its place of origin.

[4] That the boy was indeed frightened, or that a wolf was really out there in the forest is of no interest, for we are not concerned with the realist/anti-realist debate of contemporary analytic philosophy, nor with the empiricist foundation of traditional or logical positivism. Nor are we concerned with the consequences of the story – was the herd saved?, was it lost?

[5] Albert Borgmann, *Technology and the Character of Contemporary Life: A Philosophical Inquiry* (Chicago: University of Chicago Press, 1984).

[6] Jacques Ellul, *The Technological Society* (New York: Knopf, 1964; French original, 1954), pp. 3–22.

[7] The myth about the wolf, the real animal, is itself ancient; it is part Egyptian, part Greek. One variant of that myth has become a symbol and the emblem for the city of Rome, the wolf who fed human babies.

[8] For further details on the changing definitions of the terms "technology" and "science" and how they intertwine historically and etymologically, see Gayle L.

Ormiston and Raphael Sassower, *Narrative Experiments: The Discursive Authority of Science and Technology* (Minneapolis: University of Minnesota Press, 1989), chapter 1.

[9] More on the inappropriateness of using criteria for distinguishing appropriate from inappropriate technologies – including the use of the techniques of risk-assessment or risk-benefit analysis – is discussed in Ormiston and Sassower, *Narrative Experiments*, chapter 5: "Consequences of Dissemination: Narrative Recollections and the Languages of Pedagogy."

[10] Max Weber, *Economy and Society* (Berkeley: University of California Press, 1978).

[11] Thorstein Veblen's articles in the *Dial* (1919) were published as a book in 1921: *The Engineers and the Price System* (New York: Huebsch, 1921).

[12] Note Daniel Bell's introduction to Veblen's *The Engineers and the Price System* in a later edition.

[13] This view, of course, is paramount in the writings of Karl Marx and all of his successors.

[14] Martin Heidegger, "The Question Concerning Technology" (1954), in David Farrell Krell's edition of Martin Heidegger's *Basic Writings* (New York: Harper & Row, 1977), pp. 289, 308–309, 314–315.

[15] Ellul's *The Technological Society* was not translated into English until a decade after it appeared in French in 1954.

[16] Herbert Marcuse, *One-Dimensional Man: Studies in the Ideology of Advanced Industrial Society* (Boston: Beacon Press, 1964).

[17] Ellul, *The Technological Society*, p. xxxi.

[18] *Ibid.*, p. 133. See also Langdon Winner, *Autonomous Technology : Technics-Out-of-Control as a Theme in Political Thought* (Cambridge, Mass.: MIT Press, 1977).

[19] Ellul, *The Technological Society*, p. 306.

[20] Hans Jonas, *The Imperative of Responsibility: In Search of an Ethics for the Technological Age* (Chicago: University of Chicago Press, 1984), p. x.

[21] *Ibid.*, p. 28.

[22] *Ibid.*, pp. 23–24.

[23] See Mulford Q. Sibley, *Technology and Utopian Thought* (Minneapolis: Burgess, 1971).

[24] Jonas, *The Imperative of Responsibility*, pp. 144, 154, and chapter 6.

[25] *Ibid.*, pp. 203–204.

D. J. WENNEMANN

AN INTERPRETATION OF JACQUES ELLUL'S DIALECTICAL METHOD

For the past fifty years Jacques Ellul has carried out a sustained investigation and critique of the nature, development, and social implications of modern technique.[1] Much of the difficulty that has been associated with interpreting Ellul's thought[2] can be traced to a general lack of understanding of his method. It is my view that a deeper understanding of Ellul's method would show that his thought warrants serious reexamination. The purpose of this paper, then, is to attempt to describe Ellul's dialectical method.

This is an especially difficult task inasmuch as Ellul himself has said so little about his own method. In an interview with Madaleine Garrigou-Lagrange, Ellul has noted his reluctance to spell out his method for his readers in explicit terms: "I've never given an explanatory guide to my writing. I waited for readers to take the initiative and find their own explanations."[3] This attitude on the part of Ellul is, in my view, part and parcel of his general dialectical mode of thinking. Within the context of a dialectical investigation it is necessary for the reader to enter into a dynamic relationship with Ellul's texts. Their own interpretation of Ellul's thought is constructive of the body of knowledge which is to result from this dialectical process. In this regard, Ellul has noted that his readers will often recognize meanings within a text or a relational pattern between diverse works which he himself had never seen:

> In all work such as I have attempted, as in all poetic work, there is an element of inspiration which transcends the author, of which he is unaware, which is not the expression of clear ideas, and which a skilled and attentive reader can bring out, so that in listening to this reader, the author can say: "Yes, this is what I wanted to say and I did not know it." ... I have experienced this many times. No matter how willed and planned and structured my research may be, something always escapes it. There is something extra added to my something that always escapes it.[4]

In essence, the dialectic that Ellul has attempted to develop represents a totality which is really larger than its creator. It is therefore impossible for Ellul himself to be aware of all of its implications. An example of this that Ellul mentions is the occasion in which a reader pointed out to Ellul that his

Paul T. Durbin (ed.), Broad and Narrow Interpretations of Philosophy of Technology, 181–192.

work, *Hope in Time of Abandonment* (1973 [1972]), answered perfectly the problematic in *The Technological Society* (1964 [1954]). This was not a planned relationship between the works, but Ellul recognized it immediately upon its being brought to his attention.[5] Likewise, Ellul will adjust his own thought in response to his readers, both admirers and critics alike.[6] The intellectual journey, then, of a dialectician like Ellul takes on the character of a living dialogue which may grow and deepen but which can never come to a definitive conclusion. And its outcome is completely unpredictable.

Another aspect of Ellul's reluctance to define his own method is the provisional character of all knowledge within a dialectical investigation. As John Wilkinson has noted, "In Ellul's case, however, disinclination to discuss methodology specifically is almost certainly due in large part to his pervasive distrust of anything at all resembling a fixed doctrine."[7] This is confirmed by John Boli-Bennett in his study of Ellul's dialectic.[8] Dialectical methodology cannot be defined once and for all precisely because dialectic itself is a mode of thought that is constantly developing or evolving. The evolution of such a mode of thought is therefore directly opposed to systematization. Its results cannot in principle be calculated in advance. As Ellul has noted, his dialectic contains an element of inspiration which gives it the character of a living, dynamic process and not a dead mechanical procedure.

I have said that Ellul's dialectic represents a totality which is larger than its creator. This is the goal of dialectical thought in general. "The dialectician is the one who sees the totality."[9] But the peculiar thing about Ellul's dialectic is that it is made up of two distinct totalities. And this is really the result of Ellul's own personal intellectual development. In interviews Ellul had described his desire from a very early age to develop a total explanation or theory. He notes that his friend Bernard Charbonneau had done so while in his early twenties.[10]

The peculiar difficulty Ellul had in developing such a theory was that his thought was influenced by two distinct sources which seem to be antithetical to one another. They are on the one hand the works of Karl Marx, and the Christian revelation on the other. "From the beginning my thinking revolved chiefly around the contradiction between the evolution of the modern world and the Biblical content of the revelation."[11] Ellul's use of the word "contradiction" is essential here. It is because Marx and the Christian revelation both represent totalities that it became necessary for Ellul to develop a dialectical mode of thought. In Ellul's view, neither totality can be resolved into the other. If this were to be attempted it would only end in making an ideology of Christianity; it would mask the reality of the world and at the same time distort the results of a critical sociological analysis.[12] It was necessary, therefore, for Ellul to maintain both totalities and to confront the one with the other. It was in this way alone, Ellul says, that he could maintain the twofold source of his intellectual inspiration.[13]

> I realized that Christianity was a totality implying an ethic in all areas, and that Marx too claimed to be a totality. I was sometimes torn between the two extremes, and sometimes reconciled; but I absolutely refused to abandon either one. I lived my entire intellectual life in this manner. It was thus that I was progressively led to develop a mode of dialectical thinking which I constantly made my foundation.[14]

Perhaps the most significant implication of this twofold source of Ellul's thought is that the resulting dialectic cannot be resolved by abstract synthesis. Ellul saw that it is impossible to reconcile the two totalities he has discovered. And yet it was necessary to affirm the reality of both. In this situation it was necessary that Ellul's dialectic take on an existential character. The tension which exists between Marx and Christianity had to be lived through. It is not in thought alone that the contradiction is resolved but primarily in action. The alternative that Ellul was faced with was to live through the tension present in the confrontation of these two totalities or go mad.

> Under these conditions, my thinking could only develop dialectically: either I would stay rooted at the point where I was, torn in two, and would become literally schizophrenic, or I would go past the contradiction by walking, as Mao said, on both legs, being able to respond to an existential situation, to a historical or political one. When I did this, I again encountered the contradiction and I again had to go forward. This expression of Mao's – walking on two legs – referred to something entirely different, but I can't count how many times I have applied it to myself.[15]

The existential character of Ellul's dialectic has profound implications for his readers. The reader must realize that the problematic concerning technique which Ellul has developed has no simple intellectual resolution. In fact, Ellul constructs scenarios which are so rife with contradiction that there is no foreseeable escape. This requires of the reader a decision or choice in the face of an implacable reality. In a sense, for Ellul an intellectual synthesis has the function of an ideological resolution. It gives us a false sense of security. It gives us a way of justifying ourselves and of being satisfied in a life-situation that is really always ambiguous and filled with danger. With his dialectic Ellul wishes to keep alive a sense of the tension that makes for an authentic existential choice.

> The only thing that will be of any use is not synthesis or adaptation, but confrontation;

that is, bringing face to face two factors that are contradictory and irreconcilable and at the same time inescapable. For it is only out of the *decision* he makes when he experiences this contradiction – never out of adherence to an integrated system – that the Christian will arrive at a practical position.[16]

Within such an impossible situation, the usual reaction is to repress the conflict in order to make it livable. Ellul's point here is that if we make such a situation livable as it is, we effectively surrender our freedom. For, according to Ellul's conception, freedom is not a matter of adapting to our environment. It is, rather, a matter of confronting that reality and transcending it.

James Holloway has contrasted Ellul's dual orientation – correlative sociological and theological studies – with Paul Tillich's "theology of correlation." Ellul's method, however, is better designated a "theology of confrontation."[17]

The unsystematic character of Ellul's dialectic is perhaps its most striking and unsettling feature. But this feature of his thought must be seen as a direct consequence both of his purpose as a writer and of the central problematic he has identified. That problematic is to understand the relation between the rational and irrational factors in human nature, and thus to identify the field of human freedom in the modern age.[18] In this regard it is necessary to point out that dialectic is supposed to be a mode of thought that is able to follow the contours of reality, so to speak, in all of its dynamic changes. "How can one perceive the totality without perceiving that it is always changing?"[19] But the historical character of reality implies the existence of contradictory factors that subsist together and which produce a new reality.

At this point, the idea of time or history is introduced into the dialectical relation: here we have a decisive element. Often dialectic is viewed as a type of reasoning, or as the coexistence of contradictory elements. The important thing, however, is that these contradictory factors can subsist (without eliminating one another) only as they are correlative in a flow of time leading to a new situation.[20]

The significant point here is that real historical development must include the emergence of something radically new or novel. But this is to say that something completely unforeseeable is able to develop. Reality is not, in a sense, locked in, or better, closed. There is a certain irrationality about historical development. It cannot be schematized or systematized. And its outcome cannot be predicted in advance. Therefore, Ellul's dialectic is open to what lies beyond our intellectual grasp. It includes the irrationality of human history.

I thus realized that to grasp the reality of the world we live in, we must take into account the elements that cannot be included in a vast intellectual synthesis. And I completely gave up searching for either a definitive explanation of our time and of history, and for a synthesis that can encompass everything. In other words, I show in my work a systematic mind, but I have refused, you might say systematically, to produce a system.[21]

It is by developing this dialectic that Ellul is able to carry out his task of relocating humankind within the modern technological universe. What is lacking in such a universe is precisely that openness that allows for the historical development of a free subject. For technique, in Ellul's view, represents a closed world. Its development is completely mechanical. Its results are completely predictable. Ellul's dialectic itself introduces a certain negativity and thus unpredictability into this technical system. It represents a negation of technique.

In this condition it is not enough that this contradiction be that of facts, events. ... There usually has to be an express and explicit contradiction put in words by someone who contradicts, who bears the negation. In this way the negativity induces and provokes innovation, and consequently the history of the group or individual. One can see, then, that negativity has a wholly positive side. If there is transition from one state to another, we can thank negativity alone.[22]

In Ellul's view, this is precisely the task of a human being in asserting his or her subjectivity. To bear the burden of the negative is to assert one's freedom. Dialectic is thus a mode of thought that is expressive of human freedom, whether it is carried out consciously or not.[23]

What I am sure of is that life simply presupposes the innovation, that man is a history which includes negativity, and that to rest in an achieved state is in reality to deny both the life and the history or specificity of man, to claim to stop history, which is (perhaps) to enter the Kingdom of God.[24]

A fundamental principle of Ellul's dialectical method, then, is the principle of confrontation. For Ellul this involves confronting what is known sociologically of the modern world with a radically alien truth; and this is the truth of the Christian revelation.[25] John Boli-Bennett has characterized Ellul's dialectic as sociological – as opposed to the philosophical dialectic of Hegel – and the sociology Ellul draws upon is essentially that of Marx.[26] This is true as far as it goes, but it is interesting that Holloway, as I have noted, characterizes Ellul's method as a "theology of confrontation," and neither of these characterizations is completely accurate. In the end, Ellul's dialectic is a totality made up of two distinct dialectics. That is, each pole of the larger dialectic in which Ellul opposes modern sociology and theology contains its own dialectic. In this sense, Ellul's thought can be characterized as a double dialectic.

It is my view that the central theme of Ellul's thought is the reality of freedom in a technological society.[27] This represents the dialectical couplet, freedom/necessity. As we have seen, there is no synthetic moment in Ellul's dialectic, so the dialectic is in danger of being frozen at this point. It seems that the movement of thought is blocked by the opposition of these extreme positions.

It is precisely in order to break this deadlock that Ellul introduces another dialectical axis which is, in a sense, superimposed upon the original dialectical axis, somewhat like a Cartesian coordinate system – though Ellul's analysis is decidedly not a quantitative but a qualitative analysis. The freedom/necessity couplet is opposed by another dialectical couplet, the poles of which are hope and despair. This second dialectical axis, in effect, deflects the dialectical movement of the first axis, so that now Ellul can describe the despair of necessity and the despair of freedom.[28] It would be possible, in my view, although I have not seen Ellul do this explicitly, to describe the hope of necessity and the hope of freedom. To place one's hope in technique represents a false hope, one which Ellul is constantly attacking.[29] In reality, hope can only be an expression of freedom.[30]

It is this double dialectic that is peculiar to Ellul's method. The difficulty in recognizing it lies in the fact that the elements of the dialectic are dispersed in a number of Ellul's works.

At this point I would like to consider the existential character of Ellul's dialectic in terms of the theme of hope; then, finally, I will touch upon Ellul's view of humans as dialogic agents.

It is interesting that Ellul has seen in Marx's dialectic a dialectic of hope. "The fact of the matter is that Marx's thinking found such a response among the proletariat because it was a bearer of hope, and Marx's dialectic is a dialectic of hope precisely because it relates to the tangible and the concrete."[31] But it is also Ellul's view that it was not Marx who ultimately overcame Hegel's abstract dialectic, but Soren Kierkegaard. "This is where Kierkegaard's dialectic fits in. ... It is a qualitative dialectic (as contrasted with Hegel's dialectic which Kierkegaard calls quantitative), and a dialectic of life rather than a system of concepts."[32]

Vernard Eller has pointed out that while Ellul's earliest influences were Marx and Karl Barth, he identifies more strongly with the thought of Kierkegaard.[33] And it is, in his view, the existential character of Kierkegaard's thought that Ellul identifies with in particular. Certainly there is a sense in which the writings of both Kierkegaard and Ellul must be called "existential"; namely, that their discourse is life-centered rather than thought-centered, concerned with historical realities rather than with intellectual theory, and pointed toward changing behavior rather than defining doctrine.[34]

This is significant for Ellul's dialectic precisely because it is not possible to define the realities of freedom and hope. These lie in the realm of the experiential. To define them would be to undermine their historical reality since they have to do with action and not primarily with knowledge. It is for this reason that Ellul says that he does not carry out conceptual analysis.

> We are touching on a trait that I consider important: I never write ideas. I have always attempted to transmit exactly what I have experienced, in objectifying it. I have always thought on the experiential level. ... That is why my work is inevitably incomplete and does not appear to be very systematic. I have never tried to make a theoretical system concerned in itself and for itself.[35]

In a sense, Ellul conceives of this entire enterprise as flowing from the principle of confrontation. As we have seen, it is not just an intellectual enterprise Ellul has undertaken. It involves the confrontation of thought and action. It is an attempt on Ellul's part to make of the word a living reality. For Ellul, the correspondence between truth and reality is not a matter of the passive recording of events. It is, rather, a dynamic reciprocal relation between thought and action. "For the word is dialectical in itself and is at the same time integrated into the whole of existence. By this I mean that the word is intended to be lived."[36]

This emphasis on the word as a living reality is founded upon Ellul's view of human nature as having a dialogic structure. Ellul's thought can thus be placed among the dialogic existentialists, Martin Buber, Albert Camus, and Gabriel Marcel,[37] for whom dialogue is constitutive of the person. It is the dialogic character of human nature that, in Ellul's view, gives dialectical thought such an important role in maintaining human freedom. For Ellul, dialectic is at the heart of the struggle for freedom. It represents a dynamic tension that sets human subjectivity over against some object which opposes it and is thus determinate, since technique constitutes our specific environment.

David Meninger has seen this clearly in his study of Ellul's relation to Marx:

> For Ellul ... dialectic remains most valuable at the point that only provided a beginning for Marx – as a catalyst for an independent human consciousness which

may resist the weight of social constraints, as the organizer of a vision preserving human freedom.[38]

Ellul's dialectic is confrontational inasmuch as it confronts reality with both word and action. But it also confronts the reader with a discourse that cannot be grasped intellectually. Rather, it provokes a response in the form of a decision. "So I have steadily deepened this idea [of confrontation], which is meant to prompt every reader to make his own decision, on the spiritual as well as on the political or economic level."[39]

Concluding Remarks: It is my view that Ellul's thought has been misunderstood because of a lack of appreciation of the dialectical structure of his thought. Ellul's study of technique represents one pole of the dialectical axis, freedom/necessity. It represents complete determination. Technique is a system that operates by strict necessity. Reading Ellul's *The Technological Society* in isolation, one can only conclude that Ellul is a technological pessimist of the most extreme sort. But this study must be balanced by contrasting it with its theological counterpart.

Ellul has said that his plan of research was to oppose his study of technique with *The Ethics of Freedom* (1976 [1973]). In this work one discovers the absolute freedom of the Christian revelation. "We are freed only by love of God. If we place ourselves outside love, we place ourselves outside freedom. Freedom exists only in virtue of the new relation which God sets up between beings. In the old relation, which was secular and depended upon force, violence, and oppression, there could be no freedom of any kind."[40]

Ellul's intention is to attempt to make this freedom present to the technological world in which we live. In so doing, he hopes to introduce a breach in the technical system. It is Ellul's view that in this way alone are we able to live out our freedom in the deterministic technological world that we have created for ourselves.

Marquette University

NOTES

[1] Ellul defines "technique" as "the totality of means, rationally arrived at and having [as its goal] absolute efficiency (at a given stage of development) in every sphere of human endeavor." See *The Technological Society* (New York: Knopf, 1964; French original, 1954), p. xxv. It is essential to realize that Ellul is not concerned with analyzing technology. For Ellul, technology is discourse about technique, a *logos* concerning *techne*. It is on this basis that Ellul criticizes Jürgen Habermas's thought: "I am obviously in complete disagreement with Habermas (*Technik und*

Wisschenschaft als Ideologie), who confuses technology (French, 'technique') and technological discourse, and whose idea of what he means by 'Technik' strikes me as totally dated!" *The Ethics of Freedom* (Grand Rapids, Mich.: Eerdmans, 1976; French original, 1973, 1975), p. 333, n. 7.

2 See David Lovekin, "Jacques Ellul's Philosophy of Technical Consciousness," dissertation, University of Texas at Dallas, 1986. Lovekin provides a review of some of these misinterpretations.

3 *In Season, Out of Season: An Introduction to the Thought of Jacques Ellul; Based on Interviews by Madeleine Garrigou-Lagrange* (San Francisco: Harper & Row, 1982; French original, 1981), p. 73.

4 Clifford G. Christians and Jay M. van Hook, eds., *Jacques Ellul: Interpretive Essays* (Champaign: University of Illinois Press, 1981), p. 291.

5 *In Season, Out of Season*, p. 183.

6 Christians and van Hook, *Interpretive Essays*, pp. 291–292.

7 *Technological Society*, p. xiv.

8 John Boli-Bennett, "The Absolute Dialectics of Jacques Ellul," in P. Durbin, ed., *Research in Philosophy and Technology* (Greenwich, Conn.: JAI Press, 1980), p. 174.

9 Quoted in Christians and van Hook, *Interpretive Essays*, p. 293.

10 *In Season, Out of Season*, pp. 26–32.

11 Jacques Ellul, "Mirror of These Ten Years," *Christian Century*, February 18, 1970, p. 200.

12 James Y. Holloway, ed., *Introducing Jacques Ellul* (Grand Rapids, Mich.: Eerdmans, 1970), p. 6.

13 Christians and van Hook, *Interpretive Essays*, p. 305.

14 Willem H. Vanderburg, ed., *Perspectives on Our Age: Jacques Ellul Speaks on His Life and Works* (New York: Seabury Press, 1981), p. 15. Consider also this from *In Season, Out of Season*:

I thus remained unable to eliminate Marx, unable to eliminate the biblical revelation, and unable to merge the two. For me, it was impossible to put them together. So I began to be torn between the two, and I have remained so all my life. The development of my thinking can be explained starting with this confrontation (p. 16).

Also, from Christians and van Hook, *Interpretive Essays*:

To speak of "my work" would be pretentious, and yet the sum of my books constitutes a whole consciously conceived as such. At root – to the extent that I became convinced on the one side that it was impossible to make a unity of the study of modern society, and on the other that it was equally impossible to do a theological study without reference to the world in which one is – it became indispensable for me to find the link, and this could only be the dialectical process (p. 304).

15 *In Season, out of Season*, p. 18. And from Vanderburg, *Perspectives on Our Age*:

From that moment on, I lived through the conflict and the contradiction between what became the center of my life – this faith, this reference to the Bible, which I henceforth read from a different perspective – and what I knew of Marx and did not wish to abandon. For I did not see why I should have to give up the things that Marx said about society and explained about economy and injustice in the world. I saw no reason to reject them just because I was now a Christian (p. 14).

Alasdair MacIntyre's way of dealing with this conflict, in his *Marxism and Christianity* (2nd ed.; New York: Schocken, 1968), is also instructive.

[16] Ellul, "Mirror of These Ten Years" (note 11, above), p. 201. See Vernard Eller, "Ellul and Kierkegaard: Closer Than Brothers," in Christians and van Hook, *Interpretive Essays*:

On the contrary, because it deals in concrete realities rather than ideas, existential dialectic cannot arrive at synthesis, it firmly resists any effort to synthesize, and takes as its goal the finding of one's life within the tension between the dialectic's unsynthesized poles (p. 55).

[17] Holloway, *Introducing Jacques Ellul*, p. 20.

[18] Ellul, *The Technological System* (New York: Continuum, 1980; French original, 1977), p. 74:

The computer faces us squarely with the contradiction already announced throughout the technological movement and brought to its complete rigor – between the rational (problems posed because of the computer and the answers given) and the irrational (human attitudes and tendencies). The computer glaringly exposes anything irrational in a human decision, showing that a choice considered reasonable is actually emotional. It does not follow that this is translation into an absolute rationality; but plainly, this conflict introduces man into a cultural universe that is different from anything he has ever known before. Man's central, his – I might say – metaphysical problem is no longer the existence of God and his own existence in terms of that sacred mystery. The problem is now the conflict between that absolute rationality and what has hitherto constituted his person. That is the pivot of all present-day reflection, and, for a long time, it will remain the only philosophical issue.

In this way, the computer is nothing but, and nothing more than, [technique]. Yet it performs what was virtually the action of the technological whole, it brings it to its bare perfection; it makes it obvious.

It is my view that this text is a key to an adequate interpretation of Ellul's thought. This problematic can be seen in Ellul's earliest writings. The problem Ellul has identified concerning modern technique is that it has effectively altered the relationship between the rational and irrational factors in man. In a sense, technique represents a form of hyperrationality which is a result of the bifurcation of rationality from its human ground, which is constituted by various irrational drives. Today technical rationality is its own ground. This alienation of the rational and irrational factors in modern man produces a hyperrational technical development, on the one hand, and a simultaneous release of irrational drives on the other; the latter are typically expressed in political activity, a fascination with the occult, drug abuse, and sexual experimentation. The psychic disorientation we observe on such a wide scale in modern societies is a direct result of technical development. Our generalized inability to discern the operation of this technical determination upon us is a result of what Marshall McLuhan called technical anaesthesia.

[19] Christians and van Hook, *Interpretive Essays*, p. 294.

[20] *Ibid.* See also *In Season, Out of Season*: "Having said this, I want to clarify that the dialectic presupposes history. It is not enough to pose a positive factor and a negative factor. There has to be a passage of time for the two contradictory factors to come into relationship and be able to give rise to a new situation" (p. 202).
[21] *In Season, Out of Season*, p. 202.
[22] Christians and van Hook, *Interpretive Essays*, p. 296.
[23] Ellul adds, a few pages later:

My own claim, however, is that well before these intellectual formulations [of the Greeks] from the eighteenth century, B.C., dialectic appeared in Hebrew thought, and that the whole of the Old Testament expresses a dialectic. In other words, the Hebrews formulated God's revelation dialectically without examining what they were doing intellectually, without working out the noetic aspect (*Interpretive Essays*, p. 298).

[24] *Ibid.*, p. 296. See also Ellul, *The Betrayal of the West* (New York: Seabury Press, 1978): "The astounding truth that is peculiar to man: he is a maker of history. History understood as the expression of freedom and of man's mastery of events, nature, and his own life" (p. 32). Again:

Human life has no meaning if there is no chance of changing anything, no part of one's own to play, that is, if there is no history begun but not yet finished. And this is the precise moment that negativity comes to the fore. In one of my books, I adopted the famous statement of Guehenno: "The first duty of man is to say No" (Christians and van Hook, *Interpretive Essays*, p. 296).

[25] Ellul, "Mirror of These Ten Years" (note 11, above):

Consequently, I have set up the principle of confrontation. We must seek the deepest possible sociological understanding of the world we live in, apply the best methods, refrain from tampering with the results of our own research on the ground that they are "spiritually" embarrassing, maintain complete clarity and complete realism – all in order to find out, as precisely as may be, where we are and what we are doing, and also what lines of action are open to us (p. 201).

[26] Boli-Bennett, "The Absolute Dialectics of Ellul" (note 8, above), p. 174.
[27] *Technological Society*, pp. xxvi–xxxii.
[28] Ellul, "The Latest Developments in Technology and the Philosophy of the Absurd," in P. Durbin, ed., *Research in Philosophy and Technology*, vol. 7 (Greenwich, Conn.: JAI Press, 1984), pp. 92–97.
[29] Ellul, *Hope in Time of Abandonment* (New York: Seabury Press, 1973; French original, 1972).
[30] Ellul, *The Ethics of Freedom* (note 1, above).
[31] *Ibid.*, p. 87. It should be pointed out that it is Ellul's existential dialectic which has led him to posit hope as theologically primary today, rather than faith. See pp. 78–83.
[32] Ellul, *Humiliation of the Word* (Grand Rapids, Mich.: Eerdmans, 1985), p. 39. See also *Hope in Time of Abandonment* (note 29, above), pp. 53–54.
[33] Christians and van Hook, *Interpretive Essays*, p. 52.
[34] *Ibid.*, p. 57.
[35] *In Season, Out of Season*, p. 189. Ellul says:

In my case, I was able to be intellectually strict with Marx's thinking in the area of world interpretation. Moreover, I was convinced from the beginning that there could be neither Christian politics, nor Christian economics, nor a Christian society, but that the revelation contributes a fundamental existential truth. It was necessary to work it out so that these two truths could be lived together – I do mean lived, not reconciled intellectually in a system (p. 18).

36 *Humiliation of the Word* (note 32, above), p. 39.
37 See Maurice Friedman, *To Deny Our Nothingness: Contemporary Images of Man* (Delacorte Press, 1967), chapter 16: "The Existentialists of Dialogue: Marcel, Camus, Buber," pp. 281–308.
38 Christians and van Hook, *Interpretive Essays*, p. 24. Also: "Ellul's attachment to dialectic, inherited in kernel from Marx, lies at the heart of his work's purpose" (p. 29).
39 Ellul, "Mirror of These Ten Years" (note 11, above), p. 201.
40 Ellul, *Ethics of Freedom* (note 1, above), p. 248.

INDEX

PHILOSOPHY AND TECHNOLOGY

Series Editor: Paul T. Durbin

OFFICIAL PUBLICATIONS OF
THE SOCIETY FOR PHILOSOPHY AND TECHNOLOGY

---

1. *Philosophy and Technology*
   Edited by Paul T. Durbin and Friedrich Rapp.
   (Published as Volume 80 in 'Boston Studies in the Philosophy of Science')
   1983, xiv + 344pp. ISBN 90-277-1576-9
2. *Philosophy and Technology, II.* Information Technology and Computors in Theory and Practice.
   Edited by Carl Mitcham and Alois Huning.
   (Published as Volume 90 in 'Boston Studies in the Philosophy of Science')
   1986, xxii + 352pp. ISBN 90-277-1975-6
3. *Technology and Responsibility*
   Edited by Paul T. Durbin.
   1987, x + 392pp. ISBN 90-277-2415-6
4. *Technology and Contemporary Life*
   Edited by Paul T. Durbin.
   1988, viii + 320pp. ISBN 90-277-2570-5
5. *Technological Transformation.* Contextual and Conceptual Implications
   Edited by Edmund F. Byrne and Joseph C. Pitt.
   1989, xii + 314pp. ISBN 90-277-2826-7
6. *Philosophy of Technology.* Practical, Historical and Other Dimensions
   Edited by Paul T. Durbin.
   1989, xxiv + 192pp. ISBN 0-7923-0139-0
7. *Broad and Narrow Interpretations of Philosophy of Technology*
   Edited by Paul T. Durbin
   1990, xviii + 196pp. ISBN 0-7923-0684-8

Kluwer Academic Publishers
DORDRECHT / BOSTON / LONDON

MIX
Papier aus verantwortungsvollen Quellen
Paper from responsible sources
FSC® C105338
www.fsc.org

If you have any concerns about our products,
you can contact us on
ProductSafety@springernature.com

In case Publisher is established outside the EU,
the EU authorized representative is:
Springer Nature Customer Service Center GmbH
Europaplatz 3, 69115 Heidelberg, Germany

Printed by Libri Plureos GmbH
in Hamburg, Germany